Experimental Investigation of a semi-industrial Carbonate Looping Process for scale-up

TECHNISCHE UNIVERSITÄT DARMSTADT

Experimental Investigation of a semi-industrial Carbonate Looping Process for scale-up

am Fachbereich Maschinenbau
der Technischen Universität Darmstadt

zur Erlangung des Grades eines Doktor-Ingenieurs (Dr.-Ing.)
genehmigte

DISSERTATION

vorgelegt von

M.Sc. Jochen Bernd Ludwig Hilz

aus Frankenthal (Pfalz)

Erstgutachter: Prof. Dr.-Ing. Bernd Epple
Zweitgutachter: Prof. Dr.-Ing. Christian Hasse

Tag der Einreichung: 16.04.2019
Tag der mündlichen Prüfung: 12.06.2019

Darmstadt 2019
D 17

Bibliografische Information der Deutschen Nationalbibliothek

Die Deutsche Nationalbibliothek verzeichnet diese Publikation in der
Deutschen Nationalbibliografie; detaillierte bibliographische Daten sind im Internet
über http://dnb.d-nb.de abrufbar.

1. Aufl. - Göttingen: Cuvillier, 2019

Zugl.: (TU) Darmstadt, Univ., Diss., 2019

Experimental Investigation of a semi-industrial Carbonate Looping Process for scale-up
Genehmigte Dissertation von M.Sc. Jochen Bernd Ludwig Hilz aus Frankenthal (Pfalz)

Erstgutachter: Prof. Dr.-Ing. Bernd Epple
Zweitgutachter: Prof. Dr.-Ing. Christian Hasse

Tag der Einreichung: 16.04.2019
Tag der mündlichen Prüfung: 12.06.2019

Technische Universität Darmstadt - Fachbereich Maschinenbau
Darmstadt - D 17

© CUVILLIER VERLAG, Göttingen 2019

Nonnenstieg 8, 37075 Göttingen

Telefon: 0551-54724-0

Telefax: 0551-54724-21

www.cuvillier.de

ISBN 978-3-7369-7045-8

eISBN 978-3-7369-6045-9

Preface

This thesis is the from my work as a research scientist at the Institute for Energy Systems and Technology of the Department of Mechanical Engineering at the Technische Universität Darmstadt. I would like to thank Prof. Dr.-Ing. Bernd Epple, the head of the institute and supervisor of my doctoral thesis. He gave me the responsibility as project manager of the EU-funded project SCARLET and the responsibility for the semi-industrial Carbonate Looping pilot tests. With his support and promotion, he made my doctorate possible in the first place.

My special thanks also go to Prof. Dr.-Ing. Christian Hasse, head of the Institute for Simulation of Reactive Thermo-Fluid Systems, who agreed to take over the co-reporting of this work.

I would also like to thank Dr.-Ing. Jochen Ströhle, the academic councillor at the Institute for Energy Systems and Technology. He always gave me competent support in various activities such as the realization of the SCARLET project and the preparation of publications. His suggestions and tips have made a valuable contribution to the success of this work.

At this point I would like to express a heartfelt thank-you to all former research scientists at the Institute of Energy Systems and Technology who have laid an important foundation for the successful execution of this work. Furthermore, I thank my colleagues and friends Falah Alobaid, Christof Bonk, Jan-Peter Busch, Alexander Daikeler, Lorenz Frigge, Martin Haaf, Christian Heinze, Philipp Herdel, Markus Junk, Vitali Kez, David Krause, Josef Langen, Thomas Lanz, Eric Langner, Waldemar Lau, Jan May, Nicolas Mertens, Andreas Müller, Nhut Nguyen, Jens Peters, Zhichao Qu, Pascal Reinig, Michael Reitz, Ralf Starkloff, Florian Stender, Alexander Stroh, Maximilian von Bohnstein, Joachim Wagner and Coskun Yildiz. They have all contributed to the success of this work in many different ways, such as assembly in the pilot plant, during shift operation, support in planning, the preparation of reports or any other help to make this possible.

Special thanks go to my colleagues Martin Helbig and Martin Haaf, who were always open to discussions about the topics of CCS and especially Carbonate Looping, and to Susanne Tropp for the support of all my concerns and the handling of all administrative matters.

Finally, I would like to thank my parents Elke and Stefan Hilz for their unlimited support and for keeping my back free at all times. I could always count on their support and encouragement during my studies and the subsequent doctorate.

I am grateful to the European Union for the funding received from the Seventh Framework Programme for the project EU-FP7-SCARLET (Scale-up of Calcium Carbonate Looping Technology for efficient CO_2 Capture from Power and Industrial Plants, grant agreement no. 608578). The majority of the results in this work were generated within the framework of the aforementioned project.

Abstract

Carbon Capture and Storage/Utilization (CCS/U) technologies are essential for the reduction of the anthropogenic greenhouse gas emissions. CO_2 capture technologies are indispensable in the power and industry sector in order to limit the global warming to 2 °C related to the age before industrialization. The currently available CO_2 capture technologies are inevitably associated with efficiency penalties in power generation or industrial processes and high CO_2 avoidance costs. Low efficiency penalties and low additional costs for CO_2 capture are therefore of eminent importance for the further development of the CCS processes to a commercial scale and economical application. These circumstances make Carbonate Looping (CaL) an an innovative and promising post-combustion CO_2 capture technology which is mainly characterized by its low efficiency penalty, low CO_2 avoidance costs and the possibility to retrofit existing power and industrial plants.

The CaL process uses two interconnected circulating fluidized bed reactors. CO_2 contained in the flue gas from an emission source is absorbed by calcium oxide in a first reactor, the carbonator, in an exothermic reaction at around 650 °C forming calcium carbonate. The sorbent is transferred to a second reactor, the calciner. By increasing the temperature up to around 900 °C, the CO_2, bound in the solid phase, is released in an endothermic backward reaction. A gas stream of highly concentrated CO_2 leaves the calciner, while the regenerated sorbent is returned to the carbonator. The heat for the endothermic calcination reaction is provided by oxy-fuel combustion.

The present work discusses the results obtained from experimental investigation of the Carbonate Looping process in semi-industrial 1 MW_{th} scale. Long-term pilot tests were conducted to improve the process and gain reliable experimental data to scale up the CaL process to industrial size. The type of fuel, sorbent, flue gas composition, reactor design, and operating conditions were varied to assess the operability and the long-term effects on the process performance. During periods of up to 60 h, parameters were not changed to achieve steady-state conditions. The stability of the CaL process in semi-industrial size was proven by steady-state CO_2 absorption for more than 1,500 h with interconnected circulating fluidized bed reactors with absorption rates in the carbonator higher than 90 % and overall capture rates higher than 95 %.

The experimental data obtained in this work provide the basis to take the CaL process to the next stage of development, a demonstration pilot with a thermal size of 20 MW_{th}. A conceptual plant setup for this demonstration unit was derived from the operational experience in the 1 MW_{th} pilot plant. Based on evaluation of pilot testing and with the help of a validated process model, the design parameters were defined and verified. The heat and mass balancing for various operating cases was carried out to define the boundary conditions for the engineering of the demonstration pilot. Furthermore, the assessment of the demonstration pilot showed the expected performance. This in an important result to commercialize the technology in the near future.

Table of Contents

1 Introduction

The continuous increase of CO_2 concentration in the atmosphere can be observed with the beginning of industrialization in the 18th century. Since then, the CO_2 concentration has increased from pre-industrial 280 ppm to 400 ppm in 2017 [1]. The average annual increase of 0.5 ppm can be attributed to the human influence on the climate system. The current anthropogenic greenhouse gas emissions are on the highest level since the beginning of records. According to a study by the Intergovernmental Panel on Climate Change (IPCC), with a probability of over 90 % the anthropogenic greenhouse gas emissions are the main cause of climate warming and the rise in the average global temperatures. It increased by about 1 °C in the 20th century [2]. The changes have comprehensive effects on humans and nature. A drastic reduction of greenhouse gas emissions is mandatory in order to counteract it.

Combustion of fossil fuels and industrial processes count to the main sources of CO_2 emissions, others are forestry and land use [3]. Different sectors are responsible for the continuous energy-related increase of CO_2 emissions. As depicted in Fig. 1.1, the power and heat production releases almost 40 % of the energy-related CO_2 emissions. The industry and the transport sectors also have a significant share.

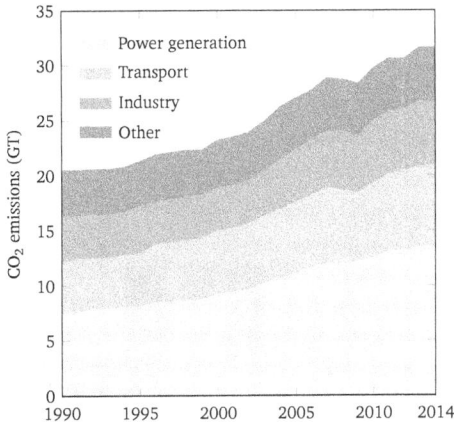

Figure 1.1: Global CO_2 emissions from 1990 to 2014 by sector [4]

At the United Nations climate conference in Paris 2015 an agreement was reached to limit global warming. The temperature increase is to be limited to 2 °C, preferably 1.5 °C above the level before industrialization. Furthermore, significant efforts needed to reduce emissions were also agreed. The currently assigned measures on national levels are not sufficient to comply with the concluded agreement [5]. The limitation of greenhouse gas emissions and thus the global warming to 2 °C requires various activities in different sections. Increased energy efficiency in industry, transport and construction sectors, shutting down of less efficient coal-fired power stations, increased investments in renewable power supply, the abolition of subsidies for fossil fuel as well as the reduction of methane emissions from oil and gas production and processing are short-term actions. Nevertheless, the achievement of long-term success to meet the climate goals demands commercial availability of advanced technologies to electrify the transport and to decarbonize the power and industry sector. These technologies will be the key to succeed in the early 2020s. However, the development of electric vehicles or CO_2 capture and storage (CCS) technologies is essential [4].

1.1 The Potential of Carbon Capture and Storage/Utilization Technologies

Carbon capture and storage (CCS) or utilization (CCU) describes the industrial intent to avoid CO_2 emissions to the atmosphere in order to meet the long-term climate goals. Therefore, the CO_2 is directly separated at the location of the emissions source, e.g. coal-fired power plants or industrial processes such as cement or steel production. The separated CO_2 allows either the storage in underground facilities (CCS) or the re-utilization (CCU), e.g. in chemical industry processes.

Since the power sector is one of the largest contributors to the overall CO_2 emissions, the International Energy Agency (IEA) considers CCS to be an essential part to reduce CO_2 emissions besides efficiency improvements, the expansion of renewables, fuel switching or nuclear power. The "New Policies Scenario" defines thereby the benchmark of the IEA and depicts today's policy ambitions. At the moment, power production emits around 500 gramme CO_2 per kilowatt-hour. In the "Sustainable Development Scenario", more than 60 % of the cumulative CO_2 emission savings occur in the power sector. This scenario outlines an integrated approach to achieve the goal of 2 ° C. According to this scenario, 17 giga-tonnes of CO_2 are saved (see Fig 1.2). CCS thereby accounts for savings of 2 giga-tonnes and is an important means to decrease the emissions to 325 gramme of CO_2 per kilowatt-hour in 2040 [6].

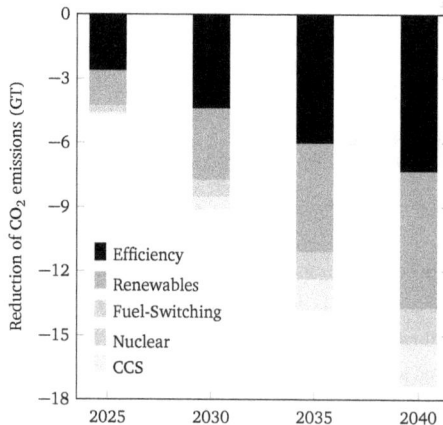

Figure 1.2: Global CO_2 emissions reduction in the "Sustainable Development Scenario" in comparison to the "New Policies Scenario" [6]

1.2 Technologies for CO$_2$ Capture

The CO$_2$ capture technologies can be divided into three processes: pre-combustion capture, post-combustion capture, and oxy-fuel combustion. The processes differ in the point where CO$_2$ is captured. It is either captured before, after or during combustion. Fig. 1.3 shows the principal scheme of the various CO$_2$ capture processes. Abanades et al. [7] and the IEAGHG [8] give a very detailed assessment of the development stages. Hereafter, a description and the development progress of first generation and second generation technologies based on fluidized beds is given.

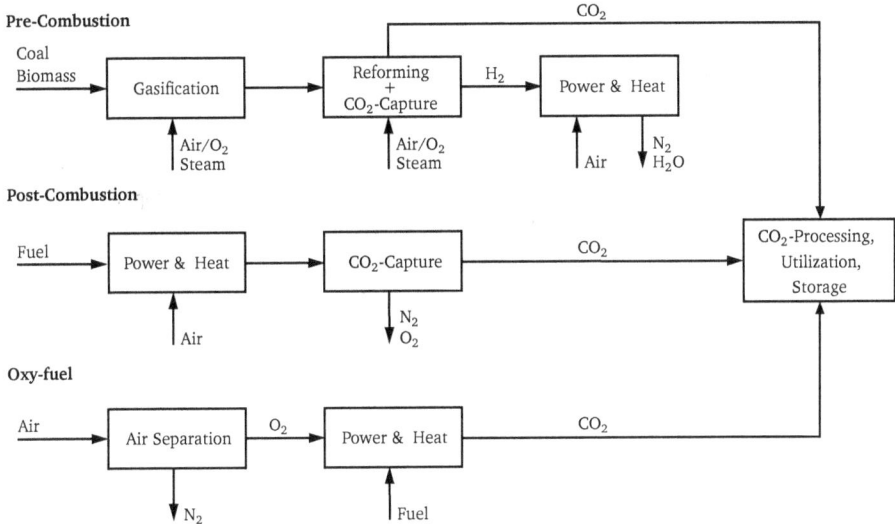

Figure 1.3: CO$_2$ capture processes [9]

The CO$_2$ capture in the pre-combustion technologies takes places before the combustion. Carbonaceous fuel is gasified or reformed. The generated product or synthesis gas mainly consists of CO$_2$ and H$_2$. The CO$_2$ is separated from this stream by gas scrubbing, e.g. with Selexol [10] or Rectisol [11]. Combustion of the remaining almost pure H$_2$ stream in a gas turbine allows heat and power production without CO$_2$ emissions. An example for a pre-combustion process is the Integrated Gasification Combined Cycle (IGCC) with CO$_2$ capture [12].

The Kemper County Energy Facility of the Mississippi Power Company has lead the way for the pre-combustion CO$_2$ capture processes. The construction of a 562 MW$_{el}$ power plant using IGCC to produce power from lignite began in 2010. The plant was to capture at least 65 % of the CO$_2$ from the synthesis gas by physical absorption (selexol scrubbing). The annually captured 3 Mio. tonnes of CO$_2$ were to be used to increase the oil production by means of Enhanced Oil Recovery (EOR). For this purpose, a pipeline network of approx. 100 km was necessary to be build for the transport of CO_2 to the neighbouring oil fields. In 2017, an increase in the costs from \$ 2.4 billion to an estimated \$ 7.5 billion and a further delay in the commercial operation of coal became apparent. The construction of the IGCC project was officially terminated. Instead, it was announced that the fuel is switched from lignite to natural gas [13].

Post-combustion CO_2 capture processes are characterised by their ability to retrofit existing emission sources such as power and industrial plants. The flue gas comes from the combustion of fossil fuel and air and is used for heat and power production. It is led into a downstream capture plant where the CO_2 is separated from the flue gas stream. The separated CO_2 then can be process and stored or utilized depending on the application. The CO_2-depleted flue gas stream of the host plant exits the capture unit and is released into atmosphere. Exemplary post-combustion processes are based on chemical absorption, such as amine scrubbing [14, 15] and chilled ammonia [16], or limestone-based absorption like Carbonate Looping [17].

Amine scrubbing has already been realized on an industrial scale with the Boundary Dam Project by the Canadian power supplier SaskPower and the Petra Nova Project by the American NRG Energy. Boundary Dam is the first commercial full-scale application of post-combustion CCS. The capture plant uses chemical absorption and separates more than 80 % of the CO_2 of a 111 MW$_{el}$ coal-fired power station. The captured 1 Mio. tonnes of CO_2 are used for EOR and the investigation of underground storage with liquid CO_2. With the beginning of capture operation in 2014, the readiness for commercial full-scale application was proven [18]. The second commercial application by Petra Nova started operation in 2017 and decarbonizes a slip stream of 37 % from the W.A. Parish Unit 8 with 654 MW$_{el}$ in Houston. The process utilizes a different amine solvent. Boundary Dam uses Shell's Cansolv solvent while Petra Nova uses Mitsubishi's KS-1 solvent. The captured CO_2 of 1.6 mega-tonnes per year is used for EOR [19].

The oxy-fuel process is based on the combustion of carbonaceous with pure oxygen [20, 21], i. e. the conversion of carbon takes places in a nitrogen-free atmosphere. Thus, the products of the combustion process are almost exclusively CO_2 and H_2O. A comparably pure stream of CO_2 exits the process after condensing the steam from the flue gas. The oxy-fuel process features the inherent CO_2 capture. The variations of oxy-fuel combustion arise by the different methods to provide the oxygen. The most common possibility is the cryogenic air separation. Other approaches are membranes processes [22] or Chemical Looping Combustion [23]. Chemical Looping Combustion provides the oxygen by alternating oxidation and reduction of a solid oxygen carrier in a high temperature process.

In 2011, the oxy-fuel technology was demonstrated by energy supplier Vattenfall in the German pilot plant Schwarze Pumpe. The CO_2 capture of more than 90 % was demonstrated with a 30 MW$_{th}$ lignite fired oxy-fuel boiler [24]. The pilot plant was built for research purposes. The amount of 1,500 tonnes of CO_2 from capture operation was sent to the storage place Ketzin. It was the first time that CO_2 captured from a power plant was stored [25]. The technological readiness ona a demonstration scale was proven with the Callide oxy-fuel project in Australia. The full chain of CCS was demonstrated by operating a 30 MW$_{el}$ oxy-fuel boiler for more than 10,000 hours in 2014 [26, 27]. But the technology has not yet been realized in a full-scale demonstration unit. A project to build a 250 MW$_{el}$ boiler in Jänschwalde in Germany has been cancelled [28].

The first generation CCS technologies like IGCC, amine scrubbing or oxy-fuel combustion have already reached demonstration or even commercial scale. Nevertheless, these technologies have the disadvantages of high energy requirements. So, they significantly decrease the efficiency of the power production from the upstream host plant or the total plant, respectively. As a consequence, the demand of energy is higher causing ecological and economic disadvantages in terms fuel consumption and increased CO_2 avoidance costs. In contrast, second generation CCS technologies such as Carbonate Looping (CaL) and Chemical Looping Combustion (CLC) offer economic and ecological improvements in terms of a better energetic efficiency and thus a lower penalty of the power production. These processes are currently under development to reach the next development level, the application in demonstration scale. Both high-temperature solid looping processes, the CaL [29, 30] and the CLC [31–33] have been a rapid development process from laboratory to successful operation in MW$_{th}$ scale. The next step will be a demonstration plant in the magnitude of order around 10-30 MW$_{th}$. Due to the unclear political conditions and the change of the power sector towards renewable energies, the required financing to bring these processes to the next stage of maturity is challenging. However, an increase of CO_2 certificates could change the conditions and may lead to a higher demand of second generation CCS technologies.

1.3 CO$_2$ Processing, Transportation, Storage and Utilization

The CCS and the CCU process chains consist of processing, transport and storage or utilization of the captured CO_2. Therefore, the CO_2 captured in the aforementioned processes has to be cleaned. Typical impurities such as H_2O, O_2, Ar, N_2, SO_x and NO_x have to be limited in the process stream since they may cause complications during transport or storage, e. g. corrosion in the pipelines. The CO_2 is processed and purified after the capture on site and the technical effort depends on the applied capture technology. The minimum requirements for the purity of CO_2 and the maximum concentrations of each species depends on the further usage. However, the components should be present in concentrations that are harmless both for the health of living organisms and with regard to technical safety [34]. Various approaches exist to condition the CO_2 product whereat most are based on multistage compression, e. g. by piston or turbo compressors, with intermediate and subsequent cooling. According to Pipitone et al. [34], the processing in general consists of the following steps:

- Compression of the CO_2 product stream to 20 bar, cooling to 25 °C and separation of water

- SO_2 removal by seawater flue gas desulphurization (SFGD)

- Compression to 33 bar and separation of water

- Adsorption of remaining water content

- Final conditioning by distillation, compression and cooling to 110 bar and 25 °C

The high pressure at ambient temperature reduces the CO_2 into a fluid state and significantly lowers the transport volume. After the CO_2 purification and compression, the transport of CO_2 captured from large point sources is feasible via pipeline or ship to the storage or utilization site. Both options allow a throughput of large amounts of CO_2 captured from commercial power and industrial plants. In general, the transport depends on the required amount and the distance between CO_2 source and sink. The on- and offshore transport via pipeline at pressures of 100-200 bar is state of the art in the USA and Canada for Enhanced Oil Recovery (EOR) where a pipeline network of 6,000 kilometers already exists [35]. The transport by ship is an economic option for distances longer than 2,000 kilometers since it is more energy-intensive and requires much more expensive infrastructure compared to pipelines. Similar to the transport of liquid natural gas, the CO_2 is liquefied to 7-9 bar at -50 °C [36].

Different possibilities for CO_2 storage (CCS) are available. Geologic formations such as deep layers of sandstone soaked by salt water (saline aquifers), depleted oil or gas fields as well as coal seams are possible storage solutions. Storage in an oil or gas field is the only commercially available technology. Especially in the USA, it is common to use CO_2 as an extraction agent into depleted oil and gas fields for Enhanced Oil (EOR) or Gas (EGR) Recovery. With respect to Europe, the capacities of saline aquifers thereby exceeds the possibilities of oil and gas reservoirs as storage options [37]. For the transport and storage a CO_2 purity of 95 vol.% is sufficient [38, 39]. The advantage is that a large part of the CO_2 injected into the storages remains there and thus is removed from the atmosphere in the long term.

An alternative to storage is the utilization of the captured CO_2 (CCU). The CO_2 can continue to be used physically or chemically. Huang and Tan [40] divides the utilization into the categories of direct use (beverages, fire extinguishers or solvents) or as a raw material for the synthesis of follow-up products. There, the required purity depending on the purpose, e. g. 99.9 vol.% in the food industry [41], has to be considered. Important follow-up products are microalgae, urea, methanol and dimethyl ether. An important factor for these products is hydrogen in sufficient quality and quantity. Hydrogen can be supplied in electrolysis driven by the power from renewable energy sources and can be utilized with CO_2 for hydration of synthetic natural gas [42]. The synthetic fuel is a possibility to store excess energy from renewable sources whereby difficulties in the storage of hydrogen could be avoided [43]. However, the development of utilization technologies requires more financial efforts to develop towards maturity and commercialization.

1.4 Costs of CO_2 Capture

The economics of the CO_2 capture processes are essential for the commercial realization. The implementation of a capture unit to thermal power plants leads to additional investment efforts, operational expenses and the efficiency of the host plant is lowered. On the one hand, the specific emissions of the power generation process decrease while on the other hand, the levelized costs of electricity generation increase. To compare the different technologies, the cost of CO_2 avoided is used as the overall cost measure. It is based on the CO_2 emission rates and the levelized costs of electricity for plants with and without CO_2 capture. The result is a quantification of average costs for avoiding a ton of atmospheric CO_2 emissions while electricity is still provided [44]. It is the basis for techno-economic comparison of the different capture technologies, but this is subject to major uncertainties. Different boundary conditions impede the scientific comparison, but the results express the relations. It is noteworthy to mention that only the specific costs for the capture are considered. Additional costs arising by transport, storage and utilization must not be taken into account to compare the technologies [45]. The real CO_2 avoidance costs can only be determined by building and operating large-scale capture units included in the whole CCS chain.

The CO_2 avoidance costs for different CO_2 capture technologies are shown in Table 1.1. It shows the results of an IEA study for first generation processes [46] and for the second generation technology CaL most recent results from the SCARLET project [47]. The depicted costs refer to the capture, processing and compression and do not include transportation and storage. The last two points mentioned are strongly dependent on the location of the site and the distance to available storage capacities.

Table 1.1: CO_2 avoidance costs of different studies for various technologies.

Technology	Study	Unit	Value
Pre-combustion	IEAGHG 2014 [46]	$€/t_{CO2}$	87
Post-combustion	IEAGHG 2014 [46]	$€/t_{CO2}$	56
Oxy-fuel combustion	IEAGHG 2014 [46]	$€/t_{CO2}$	51
Directly Heated Carbonate Looping	SCARLET [47]	$€/t_{CO2}$	23

According to Junk et al. [45], the impact factors on uncertainties in the cost evaluation are the fuel price, the operating costs, the capital costs and the hours of operation (ordered after ascending uncertainty). It can be seen in Table 1.1 that the costs of the second generation technology CaL undercuts the costs of first generation technologies. The reason for the lower costs can be explained by the significant lower efficiency penalty of the power generation by the second generation technology. In 2014, the costs given by the IEA [46] drastically increased for the first generation technologies, so that second generation CCS becomes more attractive. The costs of first generation pre-combustion (+ 321 %), post-combustion (+ 63 %) and oxy-fuel combustion (+ 56 %) technologies are estimated significantly higher compared to the results three years ago [48]. As a result, the second generation technologies, such as CaL, become much more economically attractive. Based on this knowledge, the motivation and task definition for the thesis can be derived.

1.5 Research Motivation

The objective of this work is the development of the second generation CCS technology Carbonate Looping towards the next step of maturity. The feasibility of the CaL technology has been proven in laboratory up to semi-industrial scale. The technology stands on the threshold to the next step, a demonstration plant with the size of 20 MW_{th}. To take this subsequent step, a reliable data base derived from long-term operation in semi-industrial 1 MW_{th} scale is required for scaling-up the process. In contrast to already published results, the pilot operation is carried out under realistic operating conditions of later application, e. g. coal-originated flue gas, various fuels and oxy-fuel conditions in the calciner. Additionally, the required operating periods to show the stable operability and to draw conclusion of the sorbent behaviour are several tens of hours up to days. Thus, this work shows the long-term pilot operation of the CaL technology to determine the operating parameters and conditions. It provides the reliable basis for a larger-cale demonstration plant. These fundamental expertise needed for the scale-up and pre-commercialization is obtained by addressing the key challenges and demonstration the CaL technology in 1 MW_{th} scale.

Based on the results obtained from the long-term operation 1 MW_{th} scale, a process scale-up to a 20 MW_{th} pilot plant is conducted. Developed and validated scale-up tools with the help of the experimental data support the work. A detailed process layout with heat and mass balances for a design and various operating cases is elaborated to define the basis for the industrial application in demonstration scale. The data provided by this work allows the design and engineering of reactors and all required auxiliary systems, a health, safety, environment and technical risk assessment and the estimation of investment and operational costs. Finally, it provides the confidence for investments into a larger-scale unit.

1.6 Outline of the Thesis

This thesis "Experimental Investigation of a semi-industrial Carbonate Looping Process for scale-up" is the outcome from the author's work at the Institute for Energy Systems and Technology at Technische Universität Darmstadt. The results were generated within the project SCARLET (Scale-up of Calcium Carbonate Looping Technology for efficient CO_2 Capture from Power and Industrial Plants). The thesis is structured as follows:

Chapter 2 describes the fundamentals of the Carbonate Looping process. Besides the process principles, this chapter presents the chemical and mechanical properties of the sorbent. It includes the chemical equilibrium, the kinetics of the reactions, the deactivation by impurities and the abrasion of the sorbent. In addition, the fundamentals of the fluidized bed technology are shown. This technology serves to realize the process. The most important evaluation parameters of the process are introduced and finally the progress made in the recent years and the experimental results from different CaL pilot plants worldwide are summarized.

In Chapter 3, the experimental setup for the long-term pilot tests is shown. It describes the general setup of the 1 MW_{th} pilot plant and various coupling concepts. Also, this chapter gives information about the upgrades required for long-term operation under realistic conditions. It also includes the description of the relevant measurements and the uncertainty estimation. Furthermore, the utilized sorbents and fuels during pilot operation are shown.

Chapter 4 includes the results and discussion of long-term pilot operation in $1\,MW_{th}$ scale. It describes the approach of steady-state operation and the necessary stability of operating conditions. The range of assessed parameters is given and the closure of mass and heat balances is shown. Also, the author assesses the hydrodynamic stability of various coupling concepts for the interconnection of the fluidized bed reactors and the temperature profiles. Furthermore, the chapter shows the assessment of key parameters for carbonator and calciner reactor operation and the efficiency of the process.

In Chapter 5, the CaL technology is scaled-up to a demonstration pilot of $20\,MW_{th}$ for a host plant in France. The author elaborates a detailed plant setup and scales the process on the basis of results from pilot operation with the help of a validated process model. Various operational cases besides a design case are presented to investigate the essential parameters of the process in larger scale. The scale-up includes the heat and mass balances for the design case and the ranges for the other operational cases. Finally, a detailed thermodynamic assessment is given to show the expectations for the demonstration unit.

Chapter 6 concludes the results of this work and gives an outlook for the next step of a demonstration pilot and the future application of the technology for power generation and industrial processes.

2 State of the Art of the Carbonate Looping Process

The Carbonate or Calcium Looping process (CaL) is a second generation post-combustion CO_2 capture technology. The CaL process is based on the principle of the reversible carbonation/calcination reaction. This dry sorption allows the separation of CO_2 from flue gases of power or industrial plants. Natural limestone (mainly $CaCO_3$) is utilized as sorbent for the process since it can be inexpensively mined in open-pit mining, and is available in large amounts in nature.

A system of two interconnected Circulating Fluidized Bed (CFB) reactors, a carbonator and a calciner, is applied for the CaL process. Fig. 2.1 shows the basic principle of the process.

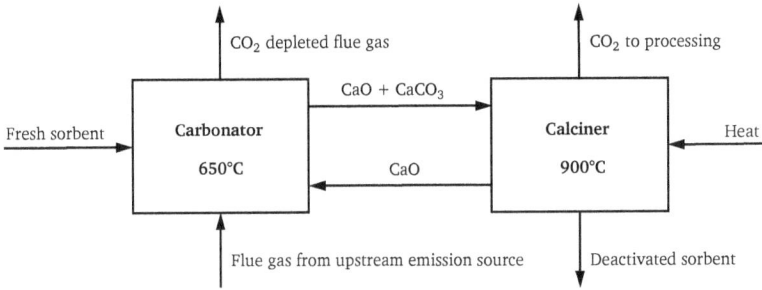

Figure 2.1: Carbonate Looping process principle.

Flue gas containing CO_2 released during combustion or a different chemical process in an upstream host plant is led through the carbonator reactor. The CO_2 is contacted with a stream of calcium oxide (CaO) particles. The gaseous CO_2 is absorbed by solid CaO in an exothermic absorption reaction (see Eq. 2.1) at optimum temperatures of around 650 °C [49] forming calcium carbonate ($CaCO_3$).

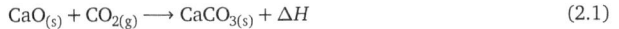

$$CaO_{(s)} + CO_{2(g)} \longrightarrow CaCO_{3(s)} + \Delta H \qquad (2.1)$$

The particle stream with the partly carbonated sorbent ($CaCO_3$) is transferred to the calciner reactor. The regeneration of the sorbent releases the CO_2 bound in the solid phase at temperatures of around 900 °C in an endothermic reaction (see Eq. 2.2) that requires additional heat input. The reaction enthalpy of carbonation (Eq. 2.1) and calcination (Eq. 2.2) is $\Delta H^0_{298K} = 178.2$ kJ/kmol.

$$CaCO_{3(s)} + \Delta H \longrightarrow CaO_{(s)} + CO_{2(g)} \qquad (2.2)$$

A highly concentrated stream of CO_2 exits the calciner and is ready for purification and compression. The regenerated sorbent is recycled to the carbonator to close the solid loop between the reactors.

The ability of the sorbent to absorb CO_2 decreases with increasing number of carbonation/calcination reaction cycles. The deactivation is mainly influenced by sintering of particle pores caused by the high process temperatures and the accumulation of impurities such as sulphur and ash. Abanades et al. [50] gives detailed investigations of the reaction kinetics of the sorbent. To maintain an certain reactivity of the sorbent, a continuous feeding of fresh sorbent (make-up) while extracting the deactivated material, is required.

2.1 Carbonate Looping Process Configurations

The development of the Carbonate Looping technology went rapidly through different stages during the last decade. Thereby, different process configurations were elaborated. The configurations can mainly be divided by the heat input in the calciner. Either the calciner is directly heated by fuel combustion or indirectly heated by a separate heat source. The directly heated CaL process for CO_2 capture from flue gases of power plants was initially proposed by Shimizu et al. [51] in 1999. Abanades et al. proposed alternative concepts for indirectly heat transfer to the calciner [52] in 2005. Both process concepts are explained below.

2.1.1 Directly Heated Carbonate Looping Process

The directly heated oxy-fired CaL process is the most developed version. The technical feasibility was proven by many pilot plants in different sizes and process configurations. The process scheme is shown in Fig. 2.2.

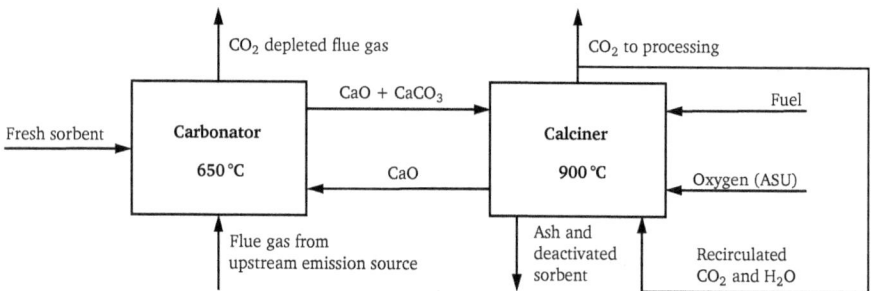

Figure 2.2: Directly heated Carbonate Looping process.

The heat input in the calciner is realized by combustion of fuel, either conventional fossil fuels such as hard coal and lignite or alternatives such as biomass and Refuse-Derived Fuel (RDF). To avoid dilution of the CO_2 product, the fuel has to be combusted under oxy-fuel conditions. Therefore, pure oxygen is diluted with recirculated flue gas and fed as an oxidant agent to the calciner. This concepts leads to some technical and energetic disadvantages.

The oxy-fuel firing of various fuels leads to technical disadvantages. Residuals from combustion enter the sorbent leading to accumulation of impurities and increased deactivation by sulphation. Sulphated material and ash represent an inert share of the circulating sorbent stream between the reactors. Due to the temperature difference between the reactors, the inert material has to be heated in the calciner and cause an increased heat demand. The inert share can be reduced by replacing the material with fresh sorbent but this also causes an increased heat demand due to the heating and first calcination of the material. Local temperature hotspots can occur caused by the high oxygen concentration for the combustion having a negative impact on the sintering of the sorbent.

The energetic disadvantage is linked to the demand of technical pure oxygen required for oxy-fuel combustion. The supply of oxygen is usually realized by cryogenic air separation. The Air Separation Unit (ASU) requires significant electric power that increases the auxiliary power. Thus, the overall efficiency of energy conversion is decreased. In addition, the ASU is related to significant investments costs.

2.1.2 Indirectly Heated Carbonate Looping Process

The Indirectly Heated CaL (IHCaL) process offers a technical and energetic optimization. A third reactor (combustor) is added to the process. This combustor provides the heat required for calcination. By operating at a higher temperature level heat can be transferred from combustor to calciner. In the combustor, fossil fuel, biomass or RDF is burned with air. Thus, an ASU is not required and the associated energetic disadvantages can be avoided. The flue gas of the combustor is led to the carbonator to be decarbonized. The process can be built and operated independently from an upstream host plant where it only decarbonizes the combustor flue gas. It also offers the possibility for a retrofit, and so the process decarbonizes the flue gas from the upstream plant and the combustor. The process scheme is shown in Fig. 2.3.

One possibility to transfer the heat from combustor to calciner is the application of heat carriers, e. g. Al_2O_3 [52], Fe_2O_3/Fe_3O_4 [53] or $CaO/CaCO_3$ [54]. Here, a material stream is circulated between calciner and combustor to utilize the sensible heat of the particles for heat transfer. However, it is technically difficult to separate the material streams of CaL and heating process. All three reactors are linked and a contamination of the sorbent with the heat carrier is unavoidable. Thus, the use of the same materials is advantageous to prevent the accumulation of impurities. Additionally, the high temperatures in the combustor promote sintering. All in all, the IHCaL process with a heat carrier is not feasible for the retrofit of existing emission sources. A possible application could be an integrated solution in the cement production process [55].

Another possibility is the indirect heat transfer from combustor to calciner by a heat exchanger, e. g. metallic walls [52] or heat pipes [56]. The application of this heat transfer concept offers some technical, energetic and economic advantages, mostly generated by the separation of combustor from the CaL process itself. A main advantage is that the process does not require an ASU. It reduces the auxiliary power and the capital investment. Further advantages are the avoidance of deactivation of the sorbent

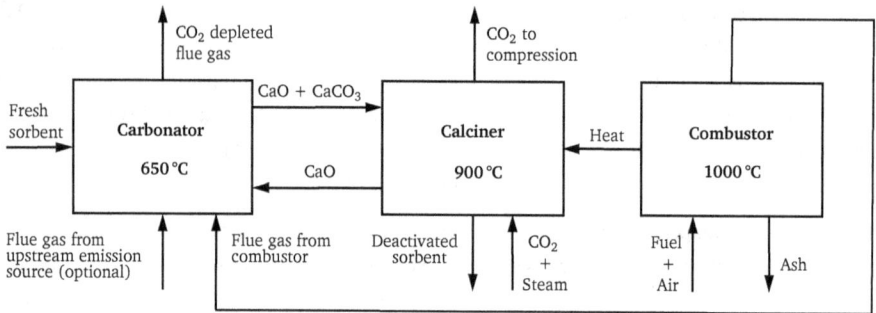

Figure 2.3: Indirectly heated Carbonate Looping process.

and the accumulation of impurities. The combustion by-products are not contacted and mixed with the sorbent. The CO_2 product is also very pure and needs no purification since only CO_2 and steam leave the calciner. In contrast, the absence of combustion gases containing water vapour creates the need for higher calcination temperatures that promotes sintering of the sorbent. To lower this effect, additional steam is optional to fluidize the calciner. This allows lower calcination temperature but adding additional costs. Furthermore, the decarbonization of flue gas from an upstream emission source leads to an significant increase of the reactor size and consequently the investment costs. From this point of view, the IHCaL process is more feasible for a greenfield application than a retrofit for existing power and industrial plants.

2.2 Fundamentals of the CaL Process

The technical fundamentals of the CaL technology that are required to understand the process and to evaluate the experimental results are presented in the following sub-sections. With respect to the behaviour of the sorbent, the following sub-sections show the temperature dependent equilibrium concentration in the CaO-$CaCO_3$ system at atmospheric pressure, various impact factors on the carbonation and calcination reaction and the effect of abrasion. The chemical properties of CaL sorbent was intensively investigated during the last decade by means of cyclic carbonation and calcination of a particle sample in a Thermogravimetric Analyzer (TGA). The attrition behaviour of the sorbent indicating the mechanical stability, was part of experiments in laboratory and pilot scale.

2.2.1 Chemical Equilibrium of CaO-CaCO₃

The carbonation/calcination cycle is an equilibrium reaction strongly dependent on the reaction temperature and the CO_2 partial pressure according to the principle of Le Chatelier. A mathematical model

2 State of the Art of the Carbonate Looping Process

for the relevant temperature range of 600-900 °C was derived by Silcox et al. [57] and serves for an estimation of the chemical equilibrium concentration (see Eq. 2.3) under atmospheric pressure [58].

$$\nu_{CO2,eq} = 4.137 \cdot 10^7 e^{\left(-\frac{20,474}{T}\right)} \tag{2.3}$$

Fig. 2.4 shows the technical relevant temperature range of 500-900 °C for Eq. 2.3. The equilibrium concentration increases with the temperature. Carbonation takes places above the equilibrium curve, below the curve the calcination reaction is predominant. An effective absorption requires low reaction temperatures to decrease the equilibrium concentration. Lowering of the carbonator temperature is limited since the reaction rate slows down. At temperatures below 500 °C, the kinetically-controlled carbonation reaction can not be maintained [59]. An equilibrium concentration of 100 % CO_2 atmosphere at atmospheric pressure is reached at 900 °C. A feasible calcination temperature depends on the process conditions and is in a range of 900-950 °C. Complete calcination at temperatures of 850-900 °C can only be realized by diluting the reaction atmosphere, e. g. with water vapour. Recent investigations confirm this possibility [58, 60].

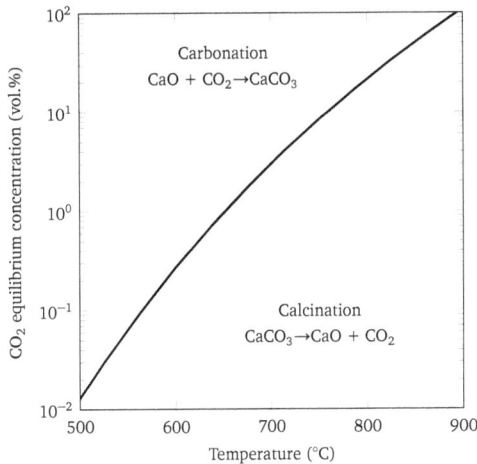

Figure 2.4: CO_2 equilibrium concentration at atmospheric pressure.

The maximum CO_2 absorption efficiency by the carbonation reaction depends on the reaction temperature and the inlet CO_2 concentration. The carbonation temperature defines the minimum CO_2 concentration at the carbonator outlet by the chemical equilibrium. This correlation is shown in Eq. 2.4.

$$E_{carb,max} = 1 - \frac{\nu_{CO2,eq}}{\nu_{CO2,carb,in}} \tag{2.4}$$

The inlet CO_2 concentration of the flue gas to be decarbonized in the CaL process depends on the type of the upstream emission source. Typical CO_2 concentrations of the flue gas stream are 4 vol.% for combined cycle power plants, 12-14 vol.% for lignite and hard coal power plants and 17 vol.% for cement plants. The maximum CO_2 absorption efficiency for these types of plants and the corresponding CO_2 inlet concentrations are shown in Fig. 2.5.

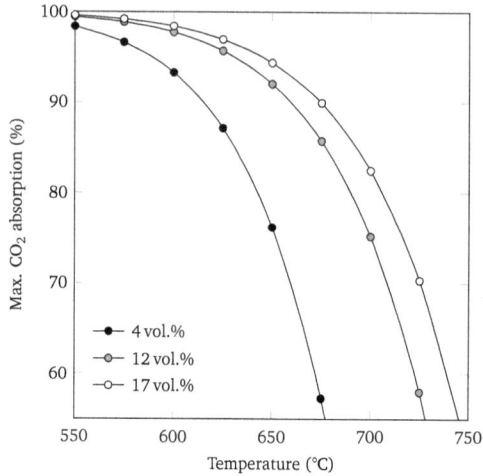

Figure 2.5: Maximum CO_2 absorption rate for various CO_2 inlet concentrations.

As shown in Fig. 2.5, the inlet CO_2 concentration defines the potential operating range of the CaL process for efficient CO_2 absorption. To achieve a carbonator absorption rate of 80 %, the maximum carbonation temperature is limited to 630 °C for combined cycle power plants since the inlet CO_2 concentration is comparably low. Decarbonization of flue gas from lignite and hard coal power plants with the same efficiency allows temperatures up to 690 °C. A decarbonization with 80 % efficiency for cement plants is possible at temperatures up to 720 °C due to the high inlet CO_2 concentration. In the lower temperature range, higher CO_2 absorption efficiencies are theoretically possible due to the equilibrium concentration. A temperature below 550 °C is not feasible since the reaction kinetics depend on the temperature, and lower temperatures slow down the reaction rate.

2.2.2 Reaction Kinetics of Carbonation/Calcination

The reaction kinetics describe the timescale of chemical reactions and physical processes. The carbonation reaction is more complex than the calcination reaction and significantly influences the CO_2 capture efficiency of the CaL process.

A main parameter to evaluate the carbonation reaction is the molar conversion or the CO_2 carrying capacity, respectively, of the sorbent. According to Sun [61], the molar conversion X is defined as the ratio of the molar mass of carbonated Ca ($CaCO_3$) and the total mass of available calcium (see Eq. 2.5).

        2 State of the Art of the Carbonate Looping Process

$$X = \frac{n_{CaCO3}}{n_{Ca}} \qquad (2.5)$$

The CO_2 absorption runs sequently in two different reaction regimes, first in a faster, kinetically-controlled one and then in a slower, diffusion-controlled reaction. In experiments up to 500 cycles of carbonation and calcination, Grasa [62] shows that this phenomenon occurs in each cycle independent of the number of cycles.

In the fast regime, CO_2 reacts with the inner particle surface. The reaction rate is high and nearly constant. It is limited by the mass transfer to the pores of the particle. A thin carbonate layer builds up on the inner particle surface and overlays or partially encloses fine pores [63] reducing the reactive surface. As soon as the carbonate layer grows until a critical thickness, the surface reaction abruptly aborts and the diffusion-controlled reaction is predominant. The CO_2 has to diffuse through the carbonate layer to react with the particle so that the reaction rate drastically decreases. The distinct reduction of the reaction rate can be explained by a low diffusivity of the carbonate layer [64]. Experiments by Alvarez et al. [65] experimentally show for numerous limestones and many different process conditions that the reaction regime changes at a critical layer thickness of 50 nm.

The molar conversion of a limestone sample in a laboratory batch-reactor over time is shown in Fig. 2.6. It is obvious that the reaction rate in the first cycle is nearly constant in the kinetically-controlled regime and passes over to diffusion-controlled one where it slowly converges to a maximum molar conversion X_{max} after a long time. The cycles 3, 5, 10 and 28 show a similar behaviour. Both reaction regimes are linearly approximated and the intersection X_{avg} represents the transition from one to the other.

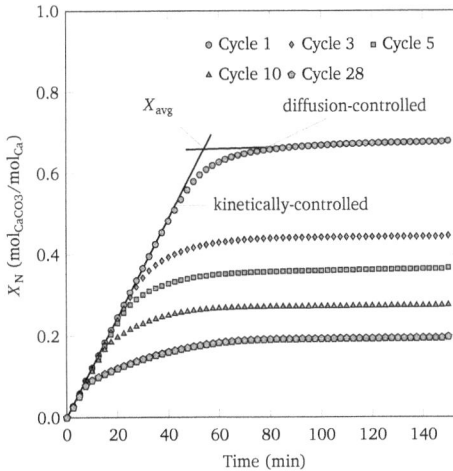

Figure 2.6: Carbonation reaction regime for different cycle numbers.

Experiments for numerous types of limestone show that the conversion mostly occurs in the kinetically-controlled regime. It is also depicted that there is a relevant discrepancy between reaching X_{avg} and the maximum molar conversion X_{max} after a long carbonation time [66]. To restrict the required residence time of the particles to carbonate, it is mostly agreed to limit the application to the kinetically-controlled fast reaction regime. More recent investigations about the carbonation reaction under steam atmosphere show that the diffusion-controlled regime also has a considerable impact on the CO_2 absorption [67].

2.2.3 Deactivation of Sorbent

The maximum carbonation conversion of natural limestone decreases with each carbonation/calcination cycle. Fig. 2.6 shows this effect for the cycle numbers 1, 3, 5, 10 and 28. This deactivation of the sorbent is understood as a reduction of the ability to absorb CO_2 or respectively as a decrease of the CO_2 transport capacity depending on the cycle number of the limestone. The decrease of molar conversion at increasing number of cycles has been described by many research groups worldwide. The decrease of CO_2 absorption ability is attributed to the continuous sintering of the pore structure of the sorbent. The morphological structure of the sorbent changes and a microscopical change from micropores to macropores takes place with the increasing number of reaction cycles. Thus, the reactive inner surface, required for the fast reaction, significantly reduces [68]. Main parameters influencing the activity of the sorbent are the temperature and the residence time, the particle size and structure as well as the reaction atmosphere during the calcination. The qualitative observations have consistently come to the same result that the decrease of activity depends on the limestone properties and process conditions.

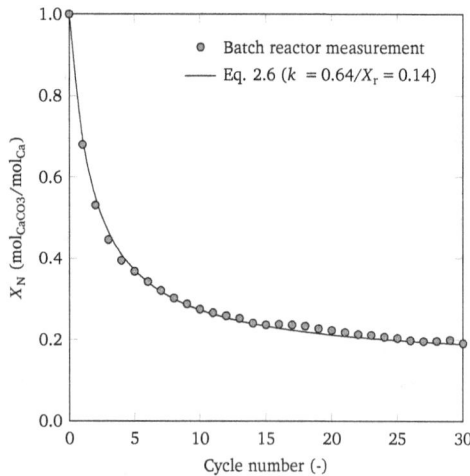

Figure 2.7: Deactivation of sorbent by cycle numbers.

The activity of limestone plays a fundamental role in the CaL process. Many models describe the behaviour of the sorbent dependent on the number of cycles. Grasa et al. [62] proposes a model based on cyclic tests with a vast number of limestones. The following model for the decrease of sorbent reactivity results from up to 500 cycles of carbonation and calcination:

$$X_N = \frac{1}{\frac{1}{1-X_r} + kN} + X_r \tag{2.6}$$

Eq. 2.6 describes the molar conversion X_N for each cycle number N wherein k is a constant for the decay of reactivity and X_r is the residual activity after an infinite number of cycles. The decrease of activity is qualitatively similar for different limestones, but there are quantitative differences in the rate of decrease in reactivity as well as the residual activity. Fig. 2.7 shows the results from batch reactor measurements of the limestone used in this work and Eq. 2.6 with the parameters $k = 0.64$ and $X_r = 0.14$. The trend shows a strong decrease of activity in the first cycles and a slow convergence towards X_r with increasing number of cycles. After just 4 cycles the activity decreased by 60 %, and after 10 cycles by 73 %.

The investigation of the sorbent deactivation have been part of the research during the last decade. On the basis of these results, the decrease of CO_2 absorption ability is attributed to the continuous sintering of the pore structure of the sorbent. The morphological structure of the sorbent changes and a microscopical change from micropores to macropores takes place with rising numbers of reaction cycles. Thus, the reactive inner surface, required for the fast reaction, significantly reduces [68]. Main parameters influencing the activity of the sorbent are the temperature and the residence time, the particle size and structure as well as the reaction atmosphere during the calcination.

2.2.3.1 Influence of Temperature and Particle Size

The sintering of the pore structure plays an important role in the decay of sorbent activity. The temperature and the residence time during the calcination cause a decrease of specific particle surface and porosity. Fig. 2.8 shows the results for an calcination exposure period of 15 minutes for a limestone particle. Temperatures below 850 °C have only a minor influence on the surface and porosity of the sorbent. However, the increase of the temperature to 950 °C leads to a significant decrease of both values. Especially for the oxy-fired CaL process, local temperatures above 850 °C are very likely so that the consequence is a possible sintering of the sorbent, especially of the micropores.

Besides the specific particle surface and the porosity, the size of the particle significantly influences the calcination rate. Effects of the particle size on the carbonation reaction have not been reported. Fig. 2.9 points out this effect. It shows the calcination rate derived from TGA analysis for a range of particle sizes from 0.4-2.0 mm. The results of this tests at 850 °C, atmospheric pressure and 0 vol.% CO_2 show the significant influence on the molar conversion of the sorbent. The calcination rate decreased with increasing particle size. The pore diffusion plays a less important role at smaller particles sizes.

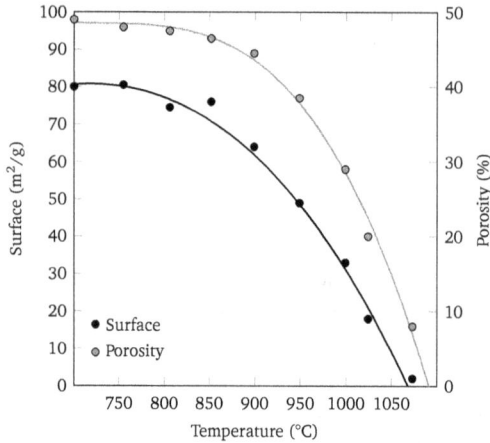

Figure 2.8: Influence of temperature on surface and porosity of limestone [69].

At small particle sizes, diffusion effects play a minor or no role for the reaction. The reaction rate is mainly dependent on the chemical reaction or the heat transfer. Hu and Scaroni [60] showed that the diffusive effects are insignificant for particles smaller than 63 μm. Since the particle size for the CaL process in CFBs is 100-400 μm, the effects of diffusion and heat transfer are relevant. Using considerably smaller particles for increasing the molar conversion and the reaction rate is not feasible with respect to energetic and technical reasons. The particles will agglomerate more, more fines will be lost via cyclones and other disruptive effects will occur in CFB systems.

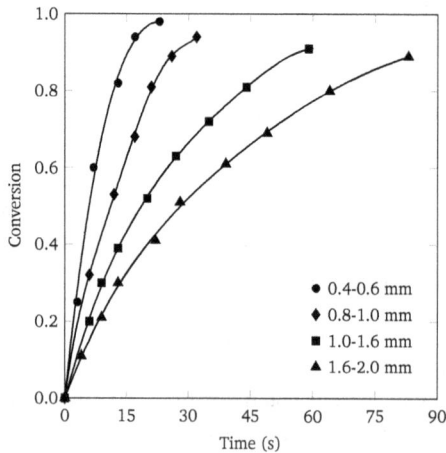

Figure 2.9: Effect of particle size on the calcination rate [58].

2.2.3.2 Influence of CO_2 Concentration

The chemical reactions of the CaO-$CaCO_3$ system depends on the CO_2 atmosphere, as indicated in Sec. 2.2.1. A higher CO_2 concentration in the carbonator helps to increase the absorption, while in the calciner the reaction is inhibited. The CO_2 concentration in the carbonator is defined by the flue gas composition of the upstream emission source to be decarbonized. However, the CO_2 concentration in the calciner can be adjusted by dilution with an easily separable gas, e.g. steam that can be condensed from the stream leaving the reactor. The investigations on the influence of the reaction rate by the CO_2 concentration were mainly focused on calcination.

Investigations in early studies by Bhatia and Perlmutter [64] cover calcination under atmospheres up to 20 vol.% CO_2 and a temperature of 910 °C. The results show that pores grow with increasing CO_2 concentration while the pore size distribution decreases. The number of the many fine pores with different sizes declines, and as a consequence, the reactive surface and thus the reaction rate significantly decreases. Particles in pure nitrogen atmosphere are more crystalline, whereas the increasing presence of CO_2 during calcination leads to more sintering.

Exemplary results of Garcia-Labiano et al. [58] are depicted in Fig. 2.10 showing the influence of the CO_2 concentration on the calcination rate at atmospheric pressure and 900 °C. The results for up to 80 vol.% CO_2 show a significant decrease of the calcination rate with increasing CO_2 partial pressure. Total conversion is then only possible with very long residence times or increased calcination temperatures. These long residence times are not feasible for a technical application in the CaL process. Thus, it is advantageous to dilute the reaction atmosphere during calcination to reduce the prevailing CO_2

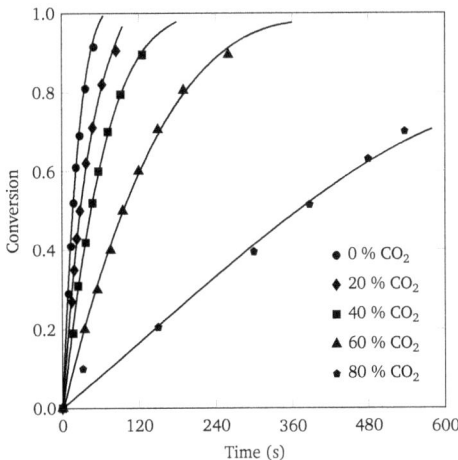

Figure 2.10: Effect of CO_2 partial pressure on the calcination rate [58].

concentration by firing a hydrogen-rich coal combined with a flue gas recirculation. Another possible solution for dilution is an additional fluidization with steam. This is not economically feasible since this option causes additional energy penalty.

2.2.3.3 Influence of Steam

The flue gas originated by combusting fuels contains steam. The typical concentration of the steam in the flue gas is 5-10 vol.% [67] and can vary for chemical processes such as steel or cement production. The steam is not only present in the carbonation of the flue gas, it is also present in the calciner as a results from fossil oxy-fuel combustion. The influence of steam on the chemical and mechanical properties of the sorbent are subject of the ongoing research.

The presence of steam during the carbonation reaction leads to a significant improvement of the sorbent. The average conversion after 10 cycles could be doubled by adding 10 % steam to the reaction atmosphere [67]. Furthermore, it has been experimentally proven how significantly the steam influences the carbonation in the different reaction regimes [70]. The main improvement occurs in the slow diffusion-controlled regime whereas the kinetically-controlled regime is not influenced. The reaction rate in the fast kinetically-controlled regime in not changed by the presence of steam [71]. According to Symonds et al. [72, 73], the presence of steam enhances the CO_2 absorption ability of the sorbent in the diffusion-controlled regime by enhancing the solid state diffusion in the product layer and significantly extends the kinetically-controlled reaction. Thus, the transition between both reaction regimes occurs at a higher molar conversion.

The influence of steam on the calcination reaction was part of several investigations that came to the conclusion that the presence of steam during calcination enhances sintering effects [67, 69, 74, 75]. Indeed, the calcination temperature can be decreased by diluting the reaction atmosphere with steam [74]. In TGA and laboratory pilot experiments it was shown that the presence of steam while decreasing the temperature of calcination leads to reduced sintering of the sorbent [76, 77]. Thus, the calcination is positively influenced by the presence of steam. It increases the reaction rate to decompose the calcium carbonate the given temperature or the required temperature to start calcination is reduced, respectively [67]. The reason for this phenomenon is the weakening of the binding between CaO and CO_2 in the calcium carbonate by the presence of steam [75].

2.2.3.4 Influence of Sulphur

Sulphur dioxide (SO_2) enters the CaL process at various points. On the one hand, the flue gas of the upstream emission source entering the carbonator could contain a certain SO_2 concentration, e. g. 500-1,000 ppm in terms of power plants. The application of flue gas desulfurization reduces the concentration to 50 ppm in order to stay within the regulatory limits. On the other hand, sulphur enters the process by the coal, containing 0-8 wt.% of sulphur [74], fired in the calciner. The presence of SO_2 leads either to indirect (see Eq. 2.7) or direct (see Eq. 2.8) sulphation of the sorbent. The predominant reaction in both reactors is the indirect sulphation since most of the solid inventory consists of CaO.

$$CaO_{(s)} + SO_{2(g)} + 0.5O_{2(g)} \longrightarrow CaSO_{4(s)} + \Delta H \tag{2.7}$$

The direct sulphation could take place in the carbonator in the case that the prevailing CO_2 concentration is higher than the equilibrium concentration, and the $CaCO_3$ does not react to CaO and CO_2. Thus, a significant effect of the direct sulphation on the process is not expected.

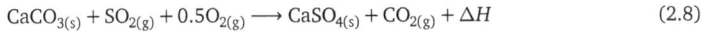

$$CaCO_{3(s)} + SO_{2(g)} + 0.5O_{2(g)} \longrightarrow CaSO_{4(s)} + CO_{2(g)} + \Delta H \tag{2.8}$$

The sulphation reaction runs in two reaction regimes, as the carbonation reaction, except the fact that a complete conversion is possible [75]. The calcium sulphate ($CaSO_4$) builds up a product layer in the pores of the sorbent particle. The dissociation of $CaSO_4$ is only possible at temperatures higher than those prevailing in the carbonate looping process [75]. The consequence is the irreversibility of sorbent sulphation leading to a loss of reactive sorbent. To maintain the carrying capacity and to prohibit the accumulation of deactivated sorbent, a continuous feed of fresh material is fed while the inactive material is purged from the process [78, 79].

The sulphation has various negative effects on the activity of thesorbent. The most apparent influence is the simultaneous reaction of carbonization and sulphation. In spite of that the carbonation reaction runs faster than sulphation, a part of the active CaO reacts with SO_2 to $CaSO_4$ and is not available for CO_2 absorption [80]. The sulphation reaction builds up an impermeable calcium sulphate layer, for fine particles on the surface and for coarser particles on the pore surface leading to a more homogeneous sulphation [74]. The $CaSO_4$ layer hinders the diffusion of CO_2 to the inner of the particle [81]. The sulphation leads to a drastic reduction of the kinetically-controlled carbonation reaction regime since the CO_2 molecules have to diffuse through the $CaSO_4$ layer right from the beginning of the reaction [82]. Due to the irreversibility of sulphation in the CaL process, the degree of sulphation grows with the increasing number of cycles [83]. Furthermore, the sulphation reaction becomes more dominant with the number of cycles [84]. As a consequence, the ability of the sorbent to absorb CO_2 decreases continuously and the $CaSO_4$ share is an inactive share of material not available for CO_2 capture. The accumulation of deactivated sorbent representing inert material leads to an increased thermal requirement in the calciner since the inert material has to be heated.

A significant sulphation of the sorbent in the CaL process is unavoidable. The good gas-solid contact promotes the reaction. The high molar Ca/S-ratio has the advantage that the most of sulphur released in the calciner combustion is directly captured, and the flue gas does not have to be desulphurized separately.

2.2.3.5 Deactivation by Impurities

The contamination of the sorbent by impurities can influence the activity of the material [85, 86]. A main role plays the reaction of the sorbent with ash entering the CaL process with the fuel combusted in the calciner. The impurities can react to many by-products, the so-called Other Calcium Compounds OCC. These include calcium silicates, particularly Ca_2SiO_2 (belite or larnite), but also dicalcium ferrite $2\,CaO{\cdot}Fe_2O_2$, calcium aluminate $CaO{\cdot}Al_2O_3$, calcium aluminosilicates $2\,CaO{\cdot}Al_2O_3{\cdot}SiO_2$ (gehlenite) and $CaO{\cdot}Al_2O_3{\cdot}2\,SiO_2$, and calcium aluminoferrites, e.g. $4\,CaO{\cdot}Al_2O_3{\cdot}Fe_2O_3$ [75]. Other calcium compounds have a much lower reactivity towards CO_2 (and SO_2) compared to CaO [87]. Thereby, the effective capacity of the sorbent to absorb CO_2 decreases, and the higher molar mass of the OCCs leads to increased sintering of the particle surface.

2.2.4 Abrasion of Sorbent

The change of particle size plays an important role in fluidized bed processes. Abrasion, agglomeration, fragmentation and chemical reactions are possible effects [88]. In CFB reactors, such as the carbonator and the calciner, occurring abrasion negatively affects the process. The particle size of the sorbent shifts from larger to smaller particles. The particle size successively decreases so that particles cannot be kept in the process, e.g. by cyclones. The consequences are procedural and economic disadvantages. The phenomena of abrasion is classified in three groups by Scala et al. [89–91]:

- Primary fragmentation: This phenomenon occurs at the first carbonization/calcination cycle and is caused by thermal stress in the particles. Rapid heating of the particles and the inner excess pressure created by the release of CO_2 in the calcination split the particles. The evoked forces decompose the particles in smaller and coarser ones. The fine particles leave the system since they cannot be retained by the cyclone.

- Secondary fragmentation: Mechanical stress on the particles induces this kind of fragmentation. Collisions with the wall and other particles crack the particles, and mainly coarser particles remain.

- Attrition by abrasion: The third mechanism describes the abrasion. The particles split due to the surface wearing caused by collisions. The result are very fine particles that are rapidly discharged from the system.

The investigations of primary fragmentation show an increased abrasion rates during the first calcination cycle [92]. Furthermore, it was shown that the breaking of particles by the release of CO_2 is stronger than the effect of the thermal shock induced by the abrupt change of temperature [93]. Higher calcination temperatures and thereby a faster reaction rate strengthens this mechanism. The abrasion behaviour worsens since higher forces act on the particle due to the increased CO_2 release [94]. The abrasion by primary fragmentation decreases with increasing number of cycles of the sorbent. With cyclic decrease of CO_2 carrying capacity, the CO_2 release in the calcination is reduced.

Secondary fragmentation and attrition phenomena mainly depend on system design. The fluidization velocity, the reactor configuration and size as well as the operational parameters are relevant parameters. Considering all relevant parameters guarantees a proper operation of the system. Another important factor to be considered is the sorbent itself. Various tests of numerous limestones show different abrasion resistances [92].

2.3 Fundamentals of Fluidized Bed Technology

The principle of fluidized beds is based on the fluidization and swirl up of solid particles. An upward flow of fluid or gas streams through a bulk of particles in a passive state. The fluid or gas then causes a liquid-like state of the particle bulk. The particles are kept in suspension. This contact process has several advantages applicable in a broad range of technical processes.

The first industrial application of fluidized bed technology in 1926 was developed by Winkler to gasify lignite for syngas production. Meanwhile, the advantages of fluidized beds in terms of good heat and mass transfer conditions are used for a large number of applications, e. g. in energy and process engineering. Fluidized beds are applied for drying, mixing, coating and crystallization processes as well as for gasification and combustion of a broad range of fuels or catalytic reactions [95].

2.3.1 Formation and Characterization of Fluidized Beds

The nature of a fluidized bed depends on the flow velocity of the fluid plus the fluid and solid properties. At low fluidization velocities the particles remain in a bulk, the so-called packed bed causing a pressure drop in the gas flow. The gas percolates through the interspace of the particles forming a packed bed. This pressure drop can be calculated according to Eq. 2.9.

$$\frac{\Delta p}{H} = 150 \frac{(1-\varepsilon)^2}{\varepsilon^3} \frac{\mu_f u}{(\phi_p d_p)^2} + 1.75 \frac{(1-\varepsilon)}{\varepsilon^3} \frac{\rho_f u^2}{\phi_p d_p} \tag{2.9}$$

The semi-empirical Eq. 2.9 of Ergun describes the bulk of solids as ordered flow channels to allow the calculation of the pressure loss. The pressure loss Δp increases with the superficial gas velocity u. Since the particles are in a passive state, only geometric size and the material properties of the fluid are considered in this equation. The solid bulk is defined by its height H and its void volume ε describing the empty space between the particles in relation to the total volume (see Eq. 2.10).

$$\varepsilon = \frac{V_{tot} - V_P}{V_{tot}} = 1 - \varepsilon_s \tag{2.10}$$

The particles in the solid bulk are swirled up by exceeding the minimum fluidization velocity u_{mf}. The minimum fluidization velocity can be calculated by the Ergun equation neglecting the inertia term in Eq. 2.9, i.e. for small Reynolds numbers.

$$u_{mf} = \frac{d_P^2(\rho_P - \rho_f)g}{150\mu_f} \frac{\varepsilon_{mf}^3 \phi_P^2}{1 - \varepsilon_{mf}}, \qquad Re_{P,mf} < 20 \tag{2.11}$$

At this point, the particles are hold in balance by the fluid and the bed shows a similar behaviour of a fluid. The viscous/drag force of the fluid exceeds the weight of the particles, i.e. the gravitational force and the sum of buoyancy and resistance are equal. Eq. 2.12 shows the balance of the forces related to the cross-sectional area of the system.

$$\Delta p = H(1 - \varepsilon)(\rho_P - \rho_f)g \tag{2.12}$$

A further increase of the superficial gas velocity leads to a uniform expansion of the bed. At higher superficial gas velocities, particles are entrained by the gas flow and discharged from the system. This characteristic velocity is called terminal velocity u_t. At the discharging point, the gravitational force and the drag force are equal. The dimensionless terminal velocity u_t^* can be calculated by a semi-empirical equation Eq. 2.13 [96].

$$u_t^* = \left[\frac{18}{(d_P^*)^2} + \frac{2.335 - 1.744\phi_P}{(d_P^*)^{1/2}} \right]^{-1} \tag{2.13}$$

The equation includes the dimensionless particle diameter d_P^* and the particle sphericity ϕ_P. The sphericity describes the roundness of the particles and is defined as the surface of a sphere relative to the surface of a particles of the same volume. By the help of Eq. 2.14, the dimensionless is converted to the dimensional one.

The characteristic flow regimes of gas/solid fluidized bed systems and their qualitative pressure loss for the superficial gas velocity are shown in Fig. 2.11. At superficial gas velocities $< u_{mf}$, the solid bulk is a packed bed (see Fig. 2.11-a). If the superficial gas velocity is increased to the range of the minimum fluidization velocity, the bed uniformly expands depending on the particle properties. The pressure loss is increased, and it is called fluidized bed (see Fig. 2.11-b). A further increase of the fluidization velocity leads to varying flow regimes that can be differentiated. First instabilities are reached when the superficial gas velocity exceeds the minimum fluidization velocity. Gas bubbles arise, and the bubble size grows and bubbles merge with increasing height of the bed. This flow regime (see Fig. 2.11-c) is called heterogeneous or Bubbling Fluidized Bed (BFB). In systems with small diameters and adequate heights, the gas bubbles can grow up to the size of the cross-section. The result is that the solid particles are lifted by the gas and then fall back. This periodically occurring phenomenon is called slugging and is accompanied by large pressure fluctuations and may cause vibration in the plant. A slugging fluidized bed is shown in Fig. 2.11-d. At superficial gas velocities $> u_t$, the solid is discharged from the system since

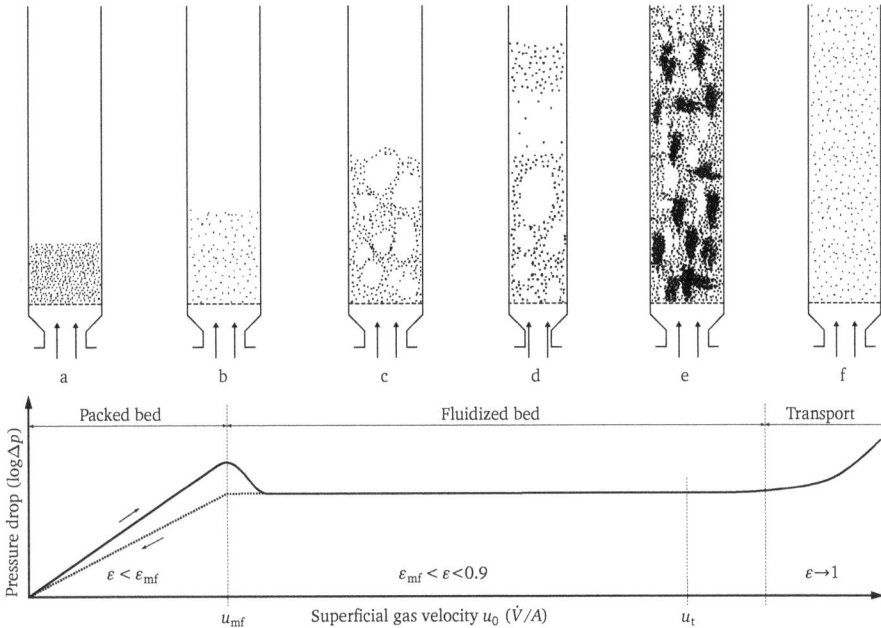

Figure 2.11: Flow regimes and pressure drop in different fluidized beds at different fluidization velocities [95].

the the fluidization velocity exceeds the terminal velocity. The initial separation between gas and solid phase vanishes and the solid concentration decreases along the reactor height. This regime is described as turbulent fluidization (see Fig. 2.11-e). Turbulent particle interactions smash the bubbles. Since a part of the solid is entrained and distracted from the system, stationary operation can be realized by equipping it with a cyclone for particle separation, a standpipe and a solid recirculation device (often designed as loop seal). This ensures a continuous material feed and is called circulating fluidized bed reactor. A further increase of the fluidization velocity leads to fast fluidization or even pneumatic transport where all particles in the system are entrained (see Fig. 2.11-f).

The solid distribution in a fluidized bed reactor at different fluidization regimes is shown in Fig. 2.12. In a bubbling fluidized bed, there is only a slight expansion induced by the slow fluidization velocity. The particles remain in the reactor except for some fine particle fractions that have a lower terminal than superficial gas velocity. The solid volume fraction ε_s in the lower part is relatively high and this is called "dense zone". The upper region with a very low solid fraction is called "lean zone". By increasing gas velocity the systems shifts to the turbulent fluidization regime, the height of the dense zone increases, and the solid volume fraction decreases there. The solid inventory is distributed more homogeneously along the reactor height. A clear dense zone is apparent and the solid fraction exponentially decreases with the height in the lean zone. In general, the conclusion can be drawn that the higher the fluidization velocity the lower is the solid volume fraction in the lower part of the reactor. At very high fluidization velocities, there is a uniform distribution of solid along the reactor height and no bed or dense zone is formed.

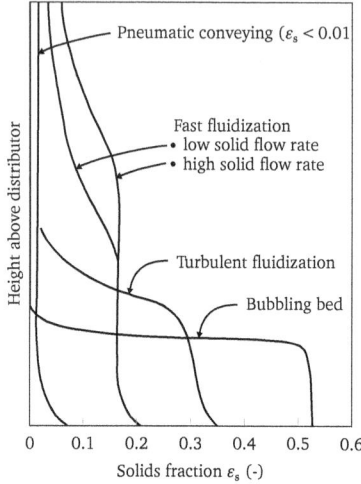

Figure 2.12: Solids distribution in fluidized bed systems at different flow regimes [97].

To design and operate fluidized bed applications, it is of great importance to know the actual fluidization regime. It plays an important role for the heat and mass transfer inside the system as well as for the constructive realization. The definition of the operational range of a fluidized bed system can be defined with the help of the state diagram for gas-solid systems (see Fig. 2.13).

With the help of the dimensionless particle diameter d_p^* on the x-axis and the dimensionless velocity u^* on the y-axis according to Eqs. 2.14 and 2.15, the operational range can be defined.

$$d_p^* = d_p \left[\frac{\rho_f(\rho_P - \rho_f)g}{\mu^2} \right]^{1/3} \tag{2.14}$$

$$u^* = u \left[\frac{\rho_f^2}{\mu(\rho_P - \rho_f)g} \right]^{1/3} \tag{2.15}$$

The carbonate looping process requires a good gas-solid contact and a continuous exchange of material between the carbonator and calciner reactor. Thus, the reactors are subsequently designed as circulating fluidized bed reactors to be operated in the turbulent or fast-fluidization regime. On the one hand, the high fluidization velocities allow very effective particle gas interactions. And on the other hand, a high solid stream is entrained from the reactor systems so that one part can be internally recycled and another

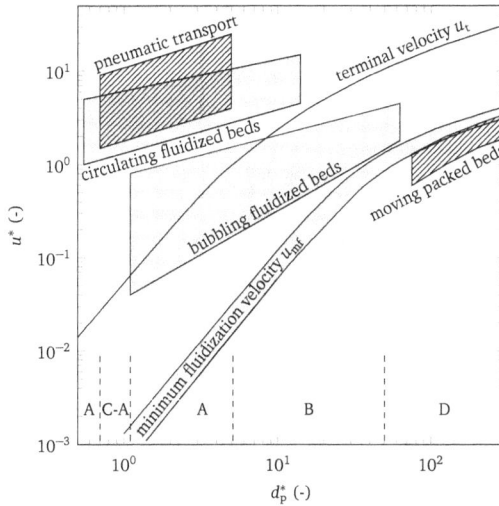

Figure 2.13: General flow regime diagram of gas-solid systems (author's illustration from Grace [98] and extended by Kunii-Levenspiel [97]).

part is transported from carbonator to calciner and vice versa. So, a continuous operation of two gas-solid reactions in the carbonate looping process can be realized with an interconnected dual circulating fluidized bed system.

2.3.2 Geldart Particle Classification

The fluidization behavior of particles in gas/solid fluidized bed systems is sustainably affected by the particle properties. Geldart [99, 100] classified the particles in four groups (A, B, C, D) according to its fluidization behaviour by means of the primary impact factors particle diameter and particle density.

- Group A, particle size 20-100 μm:
  Group A particle have small particle diameters and low particle densities $< 1,400 \, kg/m^3$. The fluidization is easily possible. The fluidized beds with these particles expand homogeneously at fluidization velocities above the minimum fluidization velocity before bubbles grow at higher velocities. Stable bubble size with a maximum diameter of 10 cm can be reached and the bubbles rise faster than the gas between the particles. After turning of the fluidization, the bed collapses just slowly. Group A particles are called "aeratable".

- Group B, particle size 40-500 μm:
  The particles of Group B have particle densities of $1,400\text{-}4,000 \, kg/m^3$. In contrast to Group A, bubble formation starts directly above the minimum fluidization velocity. The bubble size in-

---

creases with the bed height. There is no maximum bubble size because bubbles can grow due to coalescence. The bed expands moderately and collapses very fast when stopping the fluidization. The Group B particles are called "sand-like".

- Group C, particle size <30 μm:
  Group C includes very fine, cohesive particles or very fine powders. The fluidization is difficult because the adhesive forces between the particles exceed the forces can be acted on the particles by the fluidization gas. Plug flow or channel formation will occur. To realize fluidization, the fine particles can be mixed with coarser particles of the same material.

- Group D, particle size >600 μm:
  Materials of Group D have large particle sizes and/or high particle densities. Similar to group B particle, bubbles grow immediately when the minimum fluidization velocity is exceeded. The bubbles rise more slowly than the gas in the free space between the particles. Group D particles are difficult to be fluidized and the group is called "spouting".

The Geldart classification allows an easy assessment of a certain gas/solid mixture [101]. Besides particle size and particle density, other impact factors, e. g. the angular position or the roughness, influence the fluidization behaviour. The particles used for the investigation of the CaL process are mainly calcium oxide and some calcium carbonate with a particle size of 100-500 μm and a particle density of 2,700-2,800 kg/m^3. The CaL sorbent is generally related to Geldart group B (see Fig. 2.14) and appropriate for application in circulating fluidized beds.

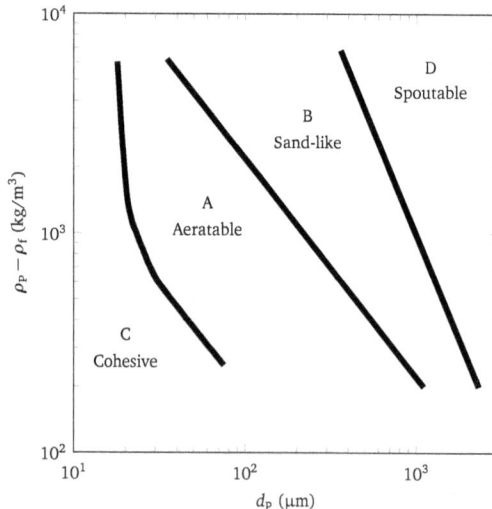

Figure 2.14: Geldart particle classification [96].

2.4 Evaluation Parameters of the Carbonate Looping Process

The evaluation of the CaL process requires the definition of evaluation parameters either directly measured or derived from various measurements. The CaL reactor system and the process scheme with the most important parameters are shown in Fig. 2.15.

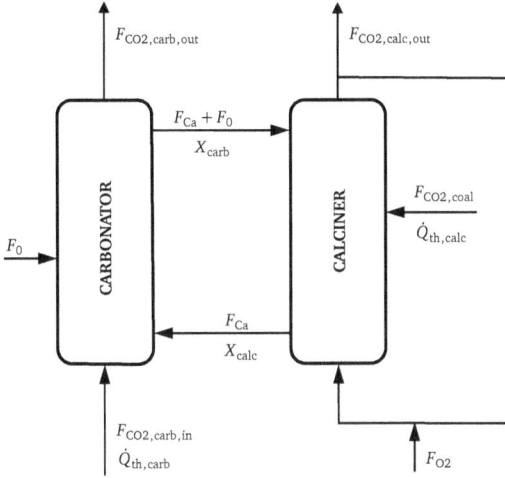

Figure 2.15: Schematics with the main parameters for CaL evaluation.

The evaluation of the directly heated CaL process can be made with the total CO_2 capture efficiency E_{tot} according to Eq. 2.16. The total efficiency is defined as the relationship between the captured CO_2 at the calciner outlet ($F_{CO2,calc,out}$) and total amount of CO_2 introduced in the process. Besides the CO_2 fed in the carbonator ($F_{CO2,carb,in}$), CO_2 formed by the oxy-fuel combustion ($F_{CO2,coal}$) and released by calcination of fresh limestone ($F_{CO2,0}$) have to be considered in the overall CO_2 balance of the process (see Fig. 2.15).

$$E_{tot} = \frac{F_{CO2,calc,out}}{F_{CO2,carb,in} + F_{CO2,0} + F_{CO2,coal}} \tag{2.16}$$

The carbonator absorption efficiency E_{carb} is one of the most crucial parameters to evaluate the quality of the process. E_{carb} is defined as the flow of CO_2 absorbed related to the flow of CO_2 fed in the carbonator from an upstream emission source and can be calculated according to Eq. 2.17.

$$E_{carb} = 1 - \frac{F_{CO2,carb,out}}{F_{CO2,carb,in}} \tag{2.17}$$

The calciner efficiency E_{calc} is of significant importance since it defines the quality and the required quantity of sorbent that has to be fed to the carbonator. Thus, E_{calc} directly influences the requirement of heat in the calciner and subsequently the operating costs of the process [102, 103]. The calciner efficiency is defined in Eq. 2.18 and considers the molar carbonate content of the circulating solid stream when entering (X_{carb}) and leaving (X_{calc}) the calciner [104].

$$E_{calc} = 1 - \frac{X_{calc}}{X_{carb}} \qquad (2.18)$$

Furthermore, the sorbent looping ratio LR is important for the evaluation. It describes the ratio of the Ca-particle stream entering the carbonator related to the incoming molar CO_2 stream to be decarbonized, as shown in Eq. 2.19. The LR mainly depends on the reactivity and the CO_2 carrying capacity, respectively. The higher the CO_2 carrying capacity, the lower LR can be. Under ideal conditions of complete molar conversion of the sorbent, the ideal looping ratio would be $LR = 1$ mole$_{Ca}$/mole$_{CO2}$ to absorb the complete stream of CO_2 entering the carbonator. Since the deactivation of the sorbent is unavoidable, the expected conversion under realistic operating conditions is significantly lower. Investigations show the need of superstoichiometry to achieve high absorption efficiencies [105]. For molar Ca conversion of 5-15 %mole$_{CaCO3}$/mole$_{Ca}$, investigations have shown required LR of 10-30 mole$_{Ca}$/mole$_{CO2}$ to achieve E_{carb} of 90 %. Increasing looping ratios lead to higher heat requirements in the calciner to bring the sorbent to the calcination temperature level. This leads to an increased oxygen demand and higher operating costs for the directly heated CaL process.

$$LR = \frac{F_{Ca}}{F_{CO2,carb,in}} \qquad (2.19)$$

Also relevant for the evaluation is the amount of make-up fed to the system. On the one hand, a certain feed is required to maintain a constant solid inventory by replacing the extracted purge in order to limit the amount of ash and sulphur accumulating [78, 79] as well as to replace the particles that leave the system via cyclone as a result of attrition. On the other hand, the make-up is required to diminish the effect of the sorbent deactivation. The fresh limestone fed as make-up (F_0) mixes with the circulating stream of sorbent to maintain and achieve a high quantity and quality of CaO. The make-up ratio MUR can be described as the ratio of the solid stream of Ca particles related to the stream of CO_2 fed to the carbonator according to Eq. 2.20 . This allows an performance-based assessment with respect to the thermal power of the upstream emission source since the incoming CO_2 flow is proportional to the load of the host plant [106].

$$MUR = \frac{F_0}{F_{CO2,carb,in}} \qquad (2.20)$$

For the application of the technology, the heat input to the calciner is an important factor in terms of technological and economical realisation. As described in Eq. 2.21, it expresses the heat input by the fuel fed to the calciner in relation to the thermal heat input equivalent of the flue gas to be decarbonized from the upstream host plant [103]. The higher the heat ratio, the higher the specific heat demand Q_s is to capture a certain amount of CO_2.

$$HR = \frac{\dot{Q}_{th,carb}}{\dot{Q}_{th,calc}} \qquad (2.21)$$

Besides the fuel input consumed for the CO_2 capture, further parasitic losses that can be evaluated are the specific oxygen ($O_{2,s}$) and the make-up consumption ($CaCO_{3,s}$) to capture a defined amount of CO_2. Electrical power is required to run the air separation unit ($\dot{Q}_{el,ASU}$) and heat from fuel input is partly used to calcine the fresh material ($\dot{Q}_{th,0}$). These parasitic consumptions can be directly expressed in relation to the total thermal capacity of the host plant and the capture unit as efficiency penalties $\Delta\eta_{ASU}$ and $\Delta\eta_0$, respectively.

$$\Delta\eta_{ASU} = \frac{\dot{Q}_{el,ASU}}{\dot{Q}_{th,carb} + \dot{Q}_{th,calc}} \tag{2.22}$$

$$\Delta\eta_0 = \frac{\dot{Q}_{th,0}}{\dot{Q}_{th,carb} + \dot{Q}_{th,calc}} \tag{2.23}$$

2.5 Overview of Carbonate Looping Pilot Testing

The synergies between the most advanced CaL process configuration of two interconnected CFB reactors and existing CFB reactor systems on an industrial scale have boosted the development of CO_2 capture by CaL. The CaL systems have rapidly developed from concept to demonstration at an industrial relevant environment (TRL 6) during the last decade [7]. Several CaL pilot plants in different scales and configurations were in operation to provide a reliable basis of experimental data. The experimental setups can be divided in two categories, laboratory scale with electrically heated reactors $< 200 kW_{th}$ and semi-industrial scale in autothermal operation $> 200\,kW_{th}$. Substantial results from both groups are described below, and the main attributes such as thermal equivalent of the decarbonized flue gas, fluidization regimes and geometries of the reactors are summarized in Table 2.1.

On the one hand, the pilot plants with a thermal size $< 200kW_{th}$ were mainly used to conduct preliminary tests and parameter variations. Generally, these rigs are not equipped with industrial system engineering, and synthetic flue gas that is used to be decarbonized. The results obtained in the laboratory are essential findings to describe and scale the process.

- The $10\,kW_{th}$ laboratory rig of Tsinghua University demonstrated CO_2 capture rates up to 95 % in intercoupled BFB reactors, a carbonator and an electrically heated calciner [107].

- A very flexible test rig is the $10\,kW_{th}$ pilot plant erected at the University of Stuttgart. The CFB or BFB carbonator can be intercoupled with an BFB or CFB calciner. The CaL process was demonstrated under oxy-firing conditions. Capture rates > 90 % by varying specific process parameters were obtained [108]. Additionally, the influence of high CO_2 concentrations and the presence of steam during on the calcination and the CO_2 capture was investigated [109].

- Consisting of an Entrained Bed (EB) carbonator and a BFB calciner, the $25\,kW_{th}$ pilot plant at Cranfield University was used to obtain a detailed sorbent composition of the sorbent. Thereby, the calciner was heated electrically and by oxy-fuel combustion with natural gas [110].

- With 30 kW$_{th}$ thermal power, the plant of CSIC-INCAR counts to the smaller units. The pilot plant consists of two intercoupled CFB reactors and the calciner is coal-fired with air. The CO_2 capture with an efficiency > 70 % was demonstrated for 8 hours [111] and the particle attrition was investigated [112]. Furthermore, a total CO_2 balance for the CO_2 capture of 70-90 % was created and the influence of the carbonator inventory on the CO_2 absorption efficiency was determined [113].

- The 75 kW$_{th}$ pilot plant of CANMET in Ottawa showed the first continuous CaL pilot operation in laboratory scale in 2008. The pilot plant consists of an Moving Bed (MB) or BFB carbonator and a CFB calciner. Pilot operation demonstrated capture rates of 70-90 % with oxygen enriched air up to 40 vol.% entering the calciner. Calciner CO_2 outlet concentration up to 83 vol.% could be reached. The sorbent from pilot operation was deeply analyzed with regard to the surface and porosity depending on the number of carbonation/calcination cycles [59, 114].

- The 120 kW$_{th}$ pilot plant at Ohio State University was erected in 2010. The carbonator is an EB reactor and the calciner an electrically heated rotary kiln with a maximum temperature of 980 °C. In addition to both reactors, the loop is extended by an hydrator to demonstrate sorbent reactivation. The results showed CO_2 capture up to 90 %, capture of SO_2 up to 99 % as well as the influence of the residence time in the EB carbonator on the performance [115].

On the other hand, pilot plants > 200kW$_{th}$ confirm the results gained in laboratory scale on a higher level. The results in semi-industrial scale prove the feasibility of the technology and confirm the long-term behaviour of the process. Especially, the sorbent deactivation requires long operation periods under realistic conditions. Thus, pilot tests need steady operation with oxy-fuel firing in the calciner and a continuous make-up feed. A homogeneous mixing of inventory by fresh limestone and the impurities fed to the process by the fuel enable the evaluation of sorbent properties with respect to further scale-up of the technology. Only the information and experience gained from pilot testing in MW scale allow a reliable re-evaluation of the process on larger scale in the light of this new experimental data.

- The CaL process was demonstrated with CO_2 capture efficiencies > 90 % in a 200 kW$_{th}$ pilot plant at the University of Stuttgart. Thereby, a Fast Fluidized Bed (FFB) or Turbulent Fluidized Bed (TFB) carbonator was operated with CFB calciner. Coal was burned with oxygen and recirculated flue gas in the calciner. The focus was on the variation of process parameters, such as sorbent looping ratio, active space time of the carbonator, in order to develop a model to describe the carbonator performance [102, 116, 117]. Additional tests with high make-up ratios were carried out to prove the feasibility of the process for application in the cement industry. The results showed a very active sorbent and the carbonator was operated close to equilibrium conditions. The increased feed of make-up led to a higher demand of coal and oxygen in the calciner [118].

- Erected in La Robla in Spain, the 300 kW$_{th}$ pilot plant of CSIC-INCAR consists of two intercoupled CFB reactors. Here, the flue gas to be decarbonized was supplied by combustion in the carbonator at 660-730 °C. The calciner is air-fired with biomass. The results showed successful co-

Table 2.1: Carbonate Looping pilot plants.

Research Institute	Size	Carbonator			Calciner		
		Type	Diameter	Height	Type	Diameter	Height/length
Tsinghua University (China)	10 kW$_{th}$	BFB	0.15 m	1 m	BFB	0.12 m	1 m
Cranefield University (United Kingdom)	25 kW$_{th}$	EB	0.1 m	4.3 m	CFB	0.17 m	1.2 m
CANMET (Canada)	75 kW$_{th}$	MB, BFB	0.1 m	4.5-5 m	CFB	0.1 m	2-5 m
Ohio State University (United States)	120 kW$_{th}$	EB	- m	- m	Rotary kiln	- m	- m
University of Stuttgart (Germany)	10 kW$_{th}$	CFB	0.07 m	12.4 m	BFB	0.1 m	3.2 m
		BFB	0.1 m	3.2 m	CFB	0.07 m	12.4 m
	200 kW$_{th}$	FFB	0.21 m	10 m	CFB	0.21 m	10 m
		TFB	0.33 m	6 m			
TU Darmstadt (Germany)	300 kW$_{th}$	CFB	0.25 m	8 m	BFB	1.1 m x 0.3 m	2.6 m
	1 MW$_{th}$	CFB	0.6 m	8.7 m	CFB	0.4 m	11.4 m
INCAR-CSIC (Spain)	30 kW$_{th}$	CFB	0.2 m	6.5 m	CFB	0.2 m	6.2 m
	300 kW$_{th}$	CFB	0.4 m	12 m	CFB	0.4 m	12 m
	1.7 MW$_{th}$	CFB	0.65 m	15 m	CFB	0.74 m	15 m
ITRI (Taiwan)	1.9 MW$_{th}$	BFB	3.3 m	0.9 m	Rotary kiln	2.75 m	5 m

EB: Entrained Bed, BFB: Bubbling Fluidized Bed, CFB: Circulating Fluidized Bed, TFB: Turbulent Fluidized Bed, FFB: Fast Fluidized Bed combustion and decarbonization of biomass in the carbonator. Long-term tests of about 60 hours with CO_2 capture rates of 60-86 % were conducted [119, 120].

- The concept of the indirectly heated CaL process was demonstrated in a $300\,kW_{th}$ pilot plant of TU Darmstadt. The system consists of a CFB carbonator and a BFB calciner and combustor, respectively. The pilot plant utilizes heat pipes [56, 121] to transfer the calcination heat from a combustor to the calciner. A long-term test of over 250 hours with CO_2 capture efficiencies of 65-90 % from synthetic flue gas proved this concept [45, 122].

- The Institute for Energy System and Technology at TU Darmstadt erected a $1\,MW_{th}$ pilot plant in 2010. This plant was the first of its kind in semi-industrial scale. The concept of CaL was proven in first test runs firing propane or coal with oxygen enriched air in the calciner. Thereby, CO_2 capture rates > 85 % were reached and the experimental data were used to validate the developed process models [123, 124]. Based on an upgraded pilot plant with flue gas recirculation in the calciner, this work shows steady-state conditions in sorbent and gas phases, the efficient operation of the carbonator and the investigation of the fuel influence under oxy-fuel calcination conditions [29, 125, 126].

- A pilot plant with a thermal size of $1.7\,MW_{th}$ was erected in La Pereda. It is connected to an existing coal-fired CFB power plant with a net electrical output of $50\,MW_{el}$. There, the CaL unit decarbonizes a slip stream of 1/150 in a dual CFB system. For 380 hours, thereof 170 hours with synthetic oxy-fuel operation in the calciner pilot tests were conducted. The temperatures of the carbonator and calciner reactor were 600-715 °C and 820-950 °C. The calciner was fluidized with an volumetric oxygen inlet concentration of 21-34 vol.% balanced with 0-75 vol.% of CO_2. Tanks supplied the gas because a recirculation of calciner flue gas is not implemented in the pilot plant. Thus, an essential part of the process was not realized since steam containing recirculated gas dilutes the CO_2 concentration in the calciner. The reaction atmosphere of calcination did not consider the presence of steam and the decreased equilibrium concentration. Additionally, the published tests were carried out without continuous make-up feed at the beginning and later only for 10 hours with continuous feed. The results did not show steady-state sorbent properties in terms of CO_2 carrying capacity and impurity accumulation with respect to long-term operation and the variation of calciner fuel. All in all, CO_2 was captured with an efficiency > 85 % and an SO_2 was captured with > 95 %, respectively [30]. Process optimizations, such as the recarbonation of sorbent to improve the carrying capacity [127] and the calciner operation under extreme oxy-fuel conditions with oxygen concentrations up 75 % [128], were tested in short-term runs. The recarbonation was tested in the carbonator loop seal at temperatures around 780 °C to increase the CO_2 carrying capacity of the sorbent. The results indicate an improvement in terms of make-up reduction but require a more accurate characterization of this novel variant. Additional tests with elevated oxygen concentration in the calciner (balanced with CO_2) were carried out and the results confirm the operability as long as sufficient solid circulation and bed inventory of solids is maintained in the calciner. For future scale-up, this process configuration could minimize or even avoid the recirculated gas for the calciner, but this approach needs to be tested further with respect to the long-term operability and the safety measures.

- The $1.9\,MW_{th}$ of the Industrial Technology Research Institute (ITRI) in Hsinchu is the largest CaL pilot worldwide and is able to capture $1.0\,t_{CO2}/h$ from the flue gas of a cement plant. The carbonator reactor is built as a BFB while the calciner is a rotary kiln fuelled with diesel under oxy-combustion conditions. This concept was demonstrated for more than 300 hours operation with CO_2 capture efficiencies $> 85\,\%$ [129, 130]

Altogether, the various pilot plants have proven the feasibility of CO_2 capture by CaL. The technology has reached TRL 6. Nevertheless, there still are technological gaps in knowledge to take it to the next level of TRL 7, and subsequently to commercial application. Crucial points to be mentioned are the decarbonization of real flue gas, the absence of flue gas recirculation in the calciner for oxy-fuel combustion and the influence of the fuel in terms of sorbent deactivation and inert material accumulation and especially the duration of test runs. Thus, the reliable assessment of CaL sorbent performance demands for long-term steady-state process operation, i. e. stable conditions of gas and sorbent phases. In particular, the mixing of existing inventory with fresh make-up material and impurities from fuel fed to the calciner influences the CO_2 carrying capacity and the accumulation of inert material. Homogeneous mixing in a steady-state condition requires long-term operation under stable conditions for several ten hours up to days. It is important to regenerate the sorbent by oxy-firing in the calciner. The recirculation of flue gas is essential since the presence of steam dilutes the CO_2 concentration in the calciner, and consequently affects the reaction kinetics and the properties of the utilized sorbent. Thus, the reliable basis of experimental data has to be provided with regard to these operating conditions for the conception of a demonstration pilot and the economic evaluation.

To bring the technology to the next step in commercialization as well as to close the gaps in operational experience and experimental results, the $1\,MW_{th}$ pilot plant was operated under realistic operating conditions for later application. In order to reach these goals, it had to be upgraded with continuous coal-originated flue gas feed from a combustor, the flue gas recirculation of the calciner reactor and improvements with respect to its operability, such as different coupling concepts between carbonator and calciner reactor. This work showed the upgraded experimental setup and assessed different reactor coupling setups for long-term operation. Based on the upgraded pilot plant, the experimental investigations of this work focussed on reaching steady-state conditions in sorbent and gas phases, the efficient operation of the carbonator and the investigation of the fuel influence under oxy-fuel calcination conditions [29, 125, 126].

3 Experimental Setup

The following chapter describes the experimental setup of the semi-industrial scale CaL pilot plant consisting of two interconnected CFB reactors and a combustion chamber with a thermal capacity of $1\,MW_{th}$ each at TU Darmstadt. The pilot was erected in 2010. First tests after the commissioning focused on batch operation [131]. Further pilot runs proved the decarbonizing of synthetic flue gas, a mixture of air and CO_2, and operating the calciner with oxygen enriched air as oxidizing agent [123, 124]. To take the process to the next level and to operate under realistic conditions for later commercial use, the pilot plant had to be upgraded. Besides the experimental setup of the upgraded pilot, the chosen sorbents and fuels are described. In addition, the applied measurements in the pilot are discussed.

3.1 $1\,MW_{th}$ pilot plant

The scheme of the upgraded $1\,MW_{th}$ CaL pilot plant is shown in Fig. 3.1. The CaL reactor system consists of a CFB carbonator with an inner diameter of $0.59\,m$ and a height of $8.6\,m$. The dimensions of the CFB calciner are $0.4\,m$ and $11\,m$, respectively. The conic calciner reactor bottom expands from $0.28\,m$ to $0.4\,m$. High efficiency cyclones at both CFB reactors separate the particles from the decarbonized flue gas and CO_2 rich stream leaving the carbonator and the calciner, respectively. The separated solids fall down in Loop Seals (LS) where a part is internally recirculated or transferred to the other reactor (from carbonator to calciner and vice versa).

Both reactors are equipped with all conventional components of industrial CFB systems, i.e. refractory lining, start-up burners fired by propane, two-pass heat exchangers to cool down the off-gases, bag filters to remove entrained solid particles, water-cooled screw conveyors to drain off solids.

A combustion chamber provides the flue gas to be decarbonized. The roof-fired furnace combusts pulverized coal with air and an induced draft fan (ID fan) keeps the combustion pressure stable. Bases on the rotating disc principle, coal is gravimetrically dosed to the furnace. The ash is removed from the reactor bottom to allow long-term operation. The flue gas to be decarbonized is then routed to the carbonator via the primary fan. Additional gases, such as CO_2, SO_2 and H_2O can be added to the fluidization gas to vary the entering concentrations to the carbonator.

To extract the released heat from the exothermic carbonation reaction and to operate at various reactor temperatures, the carbonator is equipped with five axially arranged internal cooling tubes. Thereby, the immersion depth of the lances controls the reactor temperature. There are two pairs of cooling lances and one single lance that can be moved independently from each other.

A gravimetric dosing system provides a continuous mass flow of make-up to the carbonator either replacing deactivated sorbent drained off from the reactors or compensating the loss of fines via the cyclones. The make-up stream can be fed in the standpipe or directly in the return leg of the carbonator. To automatically fill the hopper of the limestone dosing system, a pneumatic conveying system is installed.

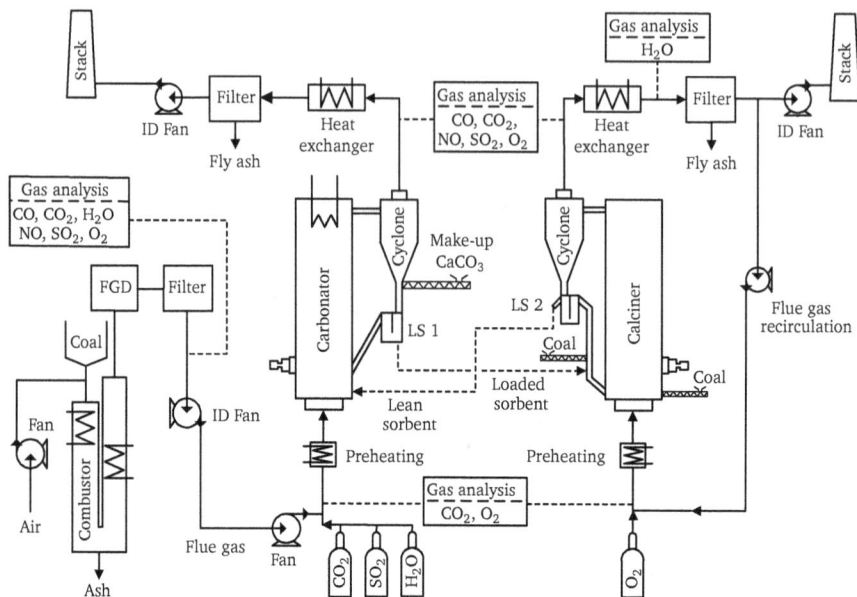

Figure 3.1: Process scheme of the 1 MW$_{th}$ CaL pilot plant.

The combustion chamber and the calcier can be fired with different types of coal, i.e. hard coal or lignite. The combustion chamber requires to be supplied with pulverized coal, whereas various particle size fractions, from pulverized to coarse sieved up to 10 mm, can be combusted in the calciner. The coal is continuously fed by a gravimetric coal dosing system either in the return leg or with a screw feeder in the reactor bottom of the calciner. The existing fuel supply of the pilot plant had to be upgraded for using lignite as fuel for the investigation of the process. The most common logistic and technical concept for supplying plants with lignite is using a silo. Hence, the pilot plant has been extended with a silo storage. The capacity of the silo is 60 m^3. A non-explosive atmosphere must be ensured in all operating conditions, which means flushing and inertization with nitrogen is required. For this purpose, the existing nitrogen supply of the pilot plant (stationary gas tank) is used. In addition, pneumatic conveying of the fuel is inert, too. Pressurized air is mixed with nitrogen to guarantee an oxygen content of the conveying fluid lower than 8 vol.%. The utilized pneumatic conveying system can both fill the hoppers of the combustion chamber and the calciner, respectively.

Calciner operation includes the possibility of air and oxy-fuel combustion. The air combustion is used for start-up to heat the reactor, whereas oxy-firing is applied in regular CaL operation. To operate the calciner under oxy-fuel conditions, a piping connection between the flue gas and the primary air duct of the calciner as well as a flue gas recirculation fan, a flow measurement device and electrical dampers are integrated. The flue gas recirculation fan serves as primary air fan if oxy-firing is not applied. The

3 Experimental Setup

oxygen required for oxy-combustion is provided by a liquefied oxygen storage tank. It is diluted with recirculated flue gas coming from the cyclone after cooling and filtration.

The operation of the CaL process requires a continuous circulation of loaded and lean sorbent between carbonator and calciner reactor. To transfer the solids between the reactors, different coupling concepts are applicable in the pilot plant. In most of the tests runs, a screw conveyor controlled the solid mass flow from carbonator to calciner. Since non-rotating systems will be applied in large-scale application, a fluidized transport valve (J-valve) was also tested during operation. The solid mass flow vice versa could be returned to the carbonator by two different concepts. In the first concept (see Fig. 3.2-a), two loop seals are used to return the solids to the carbonator. The calciner reactor is then operated in once-through without internal recirculation since the return leg is closed with a plug to avoid backflows. The entrainment in the calciner reactor balances the circulating solid flow and it is controlled by the fluidization velocity. The second carbonator to calciner coupling concept is shown in Fig. 3.2-b. The lower loop seal is removed, and a Cone Valve (CV) is installed in the upper loop seal to control the solid flow. This way, the calciner is operated with internal solid recirculation, and the superficial gas velocity is decoupled from the external solid circulation between the reactors. The solid inventory in the calciner can also be controlled by the opening of the cone valve while the fluidization is kept constant.

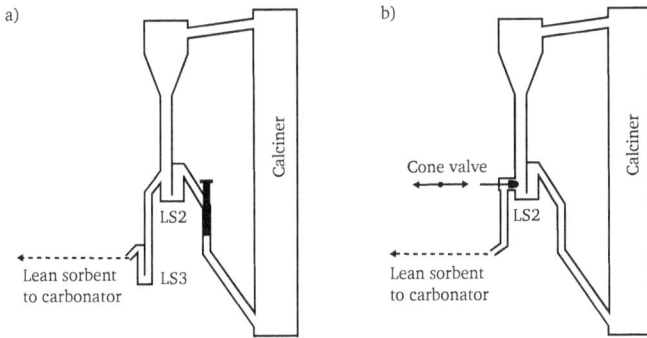

Figure 3.2: Reactor setup with two loop seals (a) and loop seal-cone valve (b) for calciner to carbonator coupling.

The plant setup is equipped with comprehensive instrumentation including temperature, pressure, flow and continuous online gas measurements installed at all relevant spots (see Sec. 3.2). Ports in the reactors provide the possibility to conduct in-bed measurements of flow patterns, i. e. particle concentration and velocity, and gas concentrations at various heights. In addition, sampling ports allow to extract material from the loop seals and reactors as well as from the heat exchangers and filters. The solid samples extracted are subsequently analysed in terms of chemical composition, CO_2 carrying capacity and particle size distribution.

The collection of all relevant measurement data from the $1\,MW_{th}$ pilot is required. For this plant, the pressures, temperatures, mass and volumetrics flows as well as the gas concentrations and the sorbent compositions are of interest. A variety of different sensors is located at characteristic and important spots in the process. The measurement data is continuously recorded and saved in the process control system. On the one hand, the main parameters are required to evaluate the process and on the other hand, many additional measurements are used to control the mass and heat flows, and to guarantee the operational safety of the plant and the personnel, respectively. Fig. 3.3 shows the CaL reactor system and the process scheme with the main parameters to be measured and determined in the pilot plant. These main variables are described as follows:

- The specific inventory of solids in the carbonator $W_{s,carb}$ and the calciner $W_{s,calc}$ is determined continuously by the measurement of the reactor pressure drop between the plane above the distributor and the top of each reactor. The pressure measurements at different reactor heights $p_{carb,i}$, $p_{calc,i}$ allow to check the vertical distribution of solids in the carbonator and calciner reactor. The Ca-based molar inventory $n_{Ca,carb}$ and $n_{Ca,calc}$ can be determined based on the chemical composition of the inventory.

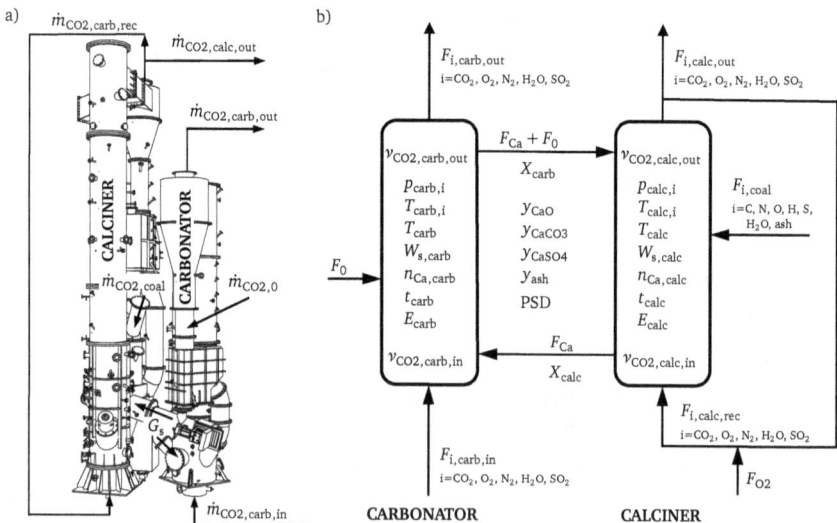

Figure 3.3: Schematics with the main mass flows (a), molar flows and operating parameters (b) of the pilot plant

- The temperature measurements at different reactor heights in the carbonator and calciner $T_{carb,i}$, $T_{calc,i}$ show the temperature profile of each reactor. The carbonator and calciner reactor temperatures are averaged by the measurements along the reactor height (T_{carb}, T_{calc}).

- The gas flows and, if necessary, their composition (CO, CO_2, H_2O, NO, O_2 and SO_2) entering and leaving the CaL system are measured. The molar flows of CO_2 ($F_{CO2,carb,in}$, $F_{CO2,carb,out}$, $F_{CO2,calc,rec}$, $F_{CO2,calc,out}$) and other species can be calculated by volumetric flow rates and the measured concentrations.

- The knowledge of mass or respectively molar flows of make-up F_0 and coal F_{coal} fed to the system is crucial to close the mass and heat balances. The composition and the Lower Heating Value (LHV) of the coal is thereby required to determine the entering species to the process, e. g. ash, and the released heat in the combustion reaction.

- Frequent analysis of the composition solids transferred from carbonator to calciner and vice versa is required. The chemical composition in terms of CaO, $CaCO_3$, $CaSO_4$ and ash content (y_{CaO}, y_{CaCO3}, y_{CaSO4}, y_{ash}) and the molar carbonate content of particles entering (X_{calc}) and leaving (X_{carb}) the carbonator can be determined. The extracted samples are analyzed for CO_2 carrying capacity and reaction rates as well as the Particle Size Distribution (PSD).

- To close the carbon balance as well as the heat balance and to determine the residence time of the particles in the reactors (t_{carb}, t_{calc}), it is crucial to determine the total circulating solid mass flow G_s and the molar circulation of Ca-based particles F_{Ca}, respectively, between carbonator and calciner reactor. It can be calculated from the total circulating solid mass flow G_s and the solid samples taken from the reactor system by Eq. 3.1, taking into account the mass fraction of Ca particles and the average molar weight of these solids.

$$F_{Ca} = \frac{G_s y_{Ca}}{M_{Ca}} = \frac{G_s(1 - y_{ash})}{M_{Ca}} \tag{3.1}$$

The applied measuring technology will be described in the following section. Since a great number of measurements is located at the peripheral systems to control and supervise the plant, such as primary air or flue gas duct or the cooling system, the focus will be on the most important measurements relevant for the evaluation of the directly heated CaL process.

3.2.1 Pressure

The pilot plant is equipped with different pressure sensors with various accuracies for the varying requirements. The measurement of the pressure inside the reactor system is of certain interest since important evaluation parameters can be derived. The solid inventory and the void fraction can be determined. The pressure profile along the reactor height shows the dense and the dilute phase. Moreover, the fluctuations of the pressure signals shows the operational state and the stability of both the fluidized bed reactors and their coupling. The pilot plant is operated near atmospheric pressure that allows the measuring of the relative pressure towards the atmosphere with high accuracy and comparably inexpensive equipment. Differential pressure measurement is only applied at positions with higher accuracy requirements, e. g. orifice plates. The positions of the pressure measurements in the CaL reactor system, required to derive and evaluate the pressure profiles, are shown in Table 3.1.

Table 3.1: Pressure measurements of the reactor system.

Component	Level above reference (m)
Carbonator	0.10, 0.22, 0.40, 0.58, 1.10, 2.07, 3.40, 8.02, 8.07
Calciner	0.91, 1.07, 1.22, 1.44, 1.65, 2.47, 2.97, 4.20, 5,10, 6.32, 8.94, 10.63
LS_{carb}	2.83, 3.83, 4.44, 6.01, 2.74
LS_{calc}	5.72, 6.57, 7.07, 8.15, 2.13
Coupling LS-LS	0.64, 1.38, 2.32, 3.96, 1.30
Coupling LS-LS-CV	5.12, 0.64, 1.38, 2.32, 3.96, 1.30
Coupling LS-CV	5.12, 2.32, 1.34

The utilized pressure transducers have a different measuring ranges of -20-20 mbar to 0-400 mbar with a measuring inaccuracy $< 0.5\%$. The pressure transducers for flow measurements in orifice plates and venturi nozzles are required to be more precise. Thus, the differential pressure transducers have a measuring inaccuracy of $< 0.1\%$ of the measuring range.

3.2.2 Temperature

The evaluation of the process demands the knowledge of the temperatures at many positions of the system. Many parameters required for the evaluation depend on the temperatures inside reactor system or on the in- and outgoing streams, respectively. The temperatures are also necessary for the calculation of heat transfer, thermodynamics or for the adherence to the operational range and plant safety.

The positions of the temperature measurements in the CaL reactor system that allow the evaluation of the temperature profiles are shown in Table 3.2. The measurement of the temperatures inside the reactor system requires thermocouples of the type N. The thermocouples have to be equipped with protective tubes of high temperature material due to a highly corrosive and erosive atmosphere in the reactors. The measuring range for the thermocouples of type N is -200-1,200°C with a maximum measuring uncertainty of 2.5 °C. For temperature measurements outside the reactor system such as the primary air or flue gas duct, Pt100 resistance thermocouples are applied. The Pt100 resistance thermocouples have a measuring error of < 0.5 °C in the range -200-1,200 °C.

Table 3.2: Temperature measurements of the reactor system.

Component	Level above reference (m)
Carbonator	0.24, 1.14, 1.55, 1.68, 2.67, 6.96, 7.75, 8.21
Calciner	1.10, 1.30, 2.10, 2.60, 4.20, 7.86, 10.55
LS_{carb}	2.78, 2.31
LS_{calc}	6.46, 2.67
Coupling LS-LS	0.55
Coupling LS-LS-CV	4.79, 0.55
Coupling LS-CV	4.79, 1.34

3.2.3 Flow

The applied flow measurements depend on the required accuracy and the economic viability of the operational purpose. Generally, all in- and outgoing streams are required for heat and mass balancing as well as for the supervision and the control of the operation. For this purpose, the data is processed and shown in the process control system.

The gas flows are mainly measured by orifice plates or venturi nozzles. The measurements include temperature and absolute pressure in front of the measuring section, and the differential pressure across it. The measured pressure and temperature values of pure gas streams are used to calculate the thermophysical properties and consequently the gas flows in real-time. Thus, the inaccuracy of orifice measurements is < 1.0 % of the upper measurement range. For the venturi nozzles or orifice plates for flue gas streams, the gas measured composition of the gas is considered in the real-time calculation. The inaccuracy for venturi nozzles is < 4.0 % of the upper measurement range due to limitations of the measurement principle. The closure of the nitrogen balance can be used to validate or reduce the relative error of the venturi measurements. Small gas flows are measured with thermal mass flow controllers with a relative error of 1.5 %. Some of the streams for flushings are measured with rotameters with an relative error of 2.5-5.0 %. The flow of the rotameters is adjusted manually and the data is not recorded in the process control system but noted for the evaluation. The measurement methods for the gas streams are shown in Table 3.3.

The solid mass flows fed to the reactor system, such as coal and make-up, are continuously detected by weighing cells. The mass flows of extracted material from filters, heat exchangers or the reactor bottoms are manually weighed and noted.

The information about the circulating solid mass flow between the reactors is of outstanding interest for the evaluation of the CaL process. The solid circulation is needed to derive main evaluation parameters

Table 3.3: Used methods for continuous measurement of the gas flow.

	Flow	Method	Rel. error in %
Carbonator inlet	Flue gas	Ring chamber orifice plate	< 1.0
	Air	Ring chamber orifice plate	< 1.0
	CO_2	Standard orifice	< 1.0
	H_2O	Standard orifice	< 1.0
	SO_2	Mass flow controller	< 1.5
Carbonator outlet	Flue gas	Venturi	< 4.0
Calciner inlet	CO_2+H_2O	Venturi	< 4.0
	O_2	Standard orifice	< 1.0
Calciner outlet	CO_2+H_2O	Venturi	< 4.0
Combustion chamber inlet	Air	Ring chamber orifice plate	< 1.0
	Natural gas	Mass flow controller	< 1.5
Combustion chamber outlet	Flue gas	Venturi	< 4.0
Carbonator loop seal	Air	Ring chamber orifice plate	< 1.0
Calciner loop seal	Air	Standard orifice	< 1.0
Coal supply	CO_2	Rotameter	2.5-5.0

such as particle residence times in the reactors, the sorbent looping ratio and to verify the heat and mass balancing. The solid circulation is also crucial for the economics with respect to scaling the technology.

Various principles allow the measurement of solid mass flows in a fluidized bed system. A continuous measurement, e.g by microwaves, is possible but difficult. The system has to be calibrated in installed state under operational conditions since the characteristics of the sorbent significantly influence the result. The relevant operational parameters affecting the result of the measurement, such as composition of the sorbent, particle size and operating temperature. These parameters change during long-term tests and make it difficult to obtain reliable values for the solid mass flow. The pilot plant has been equipped with some microwave measurement device to test the technology. Measurements spots have been installed in some part of the stand pipe that connects the loop seal of the calciner and the return leg of the carbonator. Since the technology requires a free falling of the particles the measurement spots have been positioned in a vertical part of this standpipe directly below the cone valve outlet. The solid flow can continuously be measured during operation and the qualitative change of flow conditions could be clearly indicated to increase the operability of the plant.

A more accurate determination was the application of a screw conveyor to transfer solids from carbonator to calciner. This mechanical solution can change the solid circulation by adapting the rotation speed and has linear volumetric characteristic for the operational range. Thus, the circulating solid mass flow ca be determined more accurately. This characteristic curve was first derived with fresh limestone under cold conditions. For a constant period, multiple runs at different rotations speeds were carried out and the conveyed material was weighed. The conveyed volume was determined by the help of the bulk density. During operation and for evaluation, the bulk density of the analyzed solid samples was applied with the assumption of a constant volume characteristic of the screw conveyor. This way, the conveyed mass flow can be determined for a given rotation speed with a relative error $< 10\,\%$, as shown by Helbig [132].

3.2.4 Gas Composition

The continuous recording of gas compositions of the process streams is crucial for the evaluation and the heat and mass balancing of the process. Therefore, a very small slip stream of gas is extracted from the process streams by heated gas sampling probes that are equipped with particulate filters. The particles extracted with the gas stream have to be separated to avoid errors in the subsequently arranged measuring gas conditioning. The cleaned gas is passed to the conditioning and the analyzers via heated hoses where the steam is condensed and the gas is filtered. The dry gas is routed via membrane pumps to the gas analyzers. After analysis of the gas composition, the gas is vented to a safe atmosphere. The continuous measuring methods and the maximum relative error for each species in the different process streams is shown in Table 3.4.

The determination of dry volumetric concentration for the species CO, CO_2, NO, SO_2 and CH_4 is realized by ABB URAS 206 analyzers. The principle is based on non-dispersive infrared measurement. The O_2 content is detected by paramagnetics in an ABB MAGNOS 206, H_2 by thermal conductivity in an ABB Caldos 27. The ABB gas analyzers are regularly maintained and calibrated before every test campaign. The vapour content or the H_2O fraction, respectively, is measured by Bartec Benke Hygrophil H4230-10 analyzers based on an aspiration-type psychrometer with the impinging jet method.

Table 3.4: Used methods for continuous measurement of the gas composition.

	Species	Method	Range	Unit	Rel. error in %
Carbonator inlet	CO_2	infrared	0-100	vol.%	< 0.5
	O_2	paramagnetic	0-100	vol.%	< 0.5
Carbonator outlet	CO	infrared	0-5	vol.%	< 0.5
	CO_2	infrared	0-30	vol.%	< 0.5
	O_2	paramagnetic	0-100	vol.%	< 0.5
	SO_2	infrared	0-4000	ppm	< 0.5
	NO	infrared	0-1000	ppm	< 0.5
Calciner inlet	CO_2	infrared	0-100	vol.%	< 0.5
	O_2	paramagnetic	0-100	vol.%	< 0.5

Table 3.4: Used methods for continuous measurement of the gas composition.

	Species	Method	Range	Unit	Rel. error in %
Calciner outlet	CO	infrared	0-40	vol.%	< 0.5
	CO_2	infrared	0-100	vol.%	< 0.5
	O_2	paramagnetic	0-25	vol.%	< 0.5
	SO_2	infrared	0-5	vol.%	< 0.5
	NO	infrared	0-1000	ppm	< 0.5
	H_2	paramagnetic	0-40	vol.%	< 0.5
	H_2O	psychometric	2-100	vol.%	< 1.0
	CH_4	infrared	0-5	vol.%	< 0.5
Combustion chamber	CO	infrared	0-5	vol.%	< 0.5
	CO_2	infrared	0-100	vol.%	< 0.5
	O_2	paramagnetic	0-100	vol.%	< 0.5
	SO_2	infrared	0-5	vol.%	< 0.5
	NO	infrared	0-1010	ppm	< 0.5
	H_2O	psychometric	2-100	vol.%	< 1.0

3.2.5 Sorbent Analysis

The sorbent analysis with respect to sorbent composition, particle size and reactivity is crucial to evaluate the process. These analysis were carried out by Lhoist Recherche et Developpement and Rheinkalk. The particle size was identified from analysis with an air jet sieve. The sorbent composition was derived by the loss of ignition on a NAVAS multi-sample TGA equipment. For this purpose, the residual CO_2 in the sorbent after calcination at 950 °C was determined. The analysis were completed by X-Ray Fluorescence (XRF) using a Panalytical MagiX Pro PW 2540. For selected samples, TGA analysis using a Netzsch STA 449 carbon and sulfur analysis were performed to identify the residual fuel and calcium sulphate content. The sorbent conversion and reaction kinetics were investigated in a STA 449 F3 Jupiter TGA of Netzsch and laboratory batch reactor. For selected samples, the specific surface area of the material was determined by gasadsorption with the BET method. The relative error for the analysis is 0.005-0.3 % depending on the device and the manufacturer.

3.2.6 Uncertainties

The estimation of measurement uncertainties depends on either the direct or indirect measurement principle. Physical parameters, e. g. pressure, temperature, gas concentrations are measured directly. This means that the recorded measurement value is delivered by the transmitter. In contrast, indirect measurement values are derived from various variables as the volume flow, where pressure difference and temperature of a defined constriction are used to calculate it.

All measurement transmitters have uncertainties that are usually given by the relative error. The measurement uncertainty of directly measured values depends only on the relative uncertainty of the device. For indirectly measured values or calculated parameters, the assumption of normally distributed uncertainties and the conduction of Gaussian error propagation method is conducted. Therefore, standard deviation σ of the mean value $\overline{x}$ based on multiple measurement instruments x_j can be calculated according to Eq. 3.2.

$$\sigma_{\overline{x}} = \sqrt{\sum_{j=1}^{n} \left(\frac{\partial \overline{x}}{\partial x_j} \right)^2 \sigma_{\overline{x}_j}} \tag{3.2}$$

The function of the parameter has to be differentiated based on every measurement value and multiplied with the corresponding uncertainty. The sum of the calculated uncertainties is formed and the sign is eliminated by squaring to make sure that the values do not neutralize since the uncertainties are independent from each other. The determined uncertainties of the relevant 1 MW$_{th}$ pilot plant measurements are shown in Table 3.5.

Table 3.5: Estimation of relative measurement uncertainty for the 1 MW$_{th}$ pilot plant.

Parameter	Measurement principle	Unit	Uncertainty in %
Temperature	Thermocouples, resistance thermometers	°C	< 1.0
Pressure	Transmitter with capacitive membrane	mbar	< 1.0
Coal feed	Weighing cells	kg/h	< 2.0
Make-up feed	Weighing cells	kg/h	< 1.0
Gas concentration	Infrared, paramagnetic, heat conduction	vol.%	< 1.0
Moisture	Impinging jet psychrometer	vol.%	< 1.0
Air	Ring chamber orifice plate	Nm^3/h	< 1.0
Air, CO_2, H_2O	Standard orifice	Nm^3/h	< 1.0
Flue gases	Venturi nozzle	Nm^3/h	< 4.0
C_3H_8, SO_2	Mass flow controller	Nm^3/h	< 1.5
E_{carb}	Calculation by several parameters	%	< 1.2
E_{calc}	Calculation by several parameters	%	< 5.0
LR	Calculation by several parameters	%	< 5.0
MUR	Calculation by several parameters	%	< 5.0
HR	Calculation by several parameters	%	< 5.0

Since most parameters are measured directly, the measurement uncertainties are kept within reasonable bounds. In general, the directly measured values are below 2.0%. The indirectly measured values are kept below 5.0%. The estimation of the uncertainties of MUR, LR, HR and E_{calc} is not possible with the Gaussian error propagation since there the main uncertainty occurs in the collection of the sorbent and coal samples and not in the analysis. Estimation of the uncertainty of the sampling procedure requires a more precise examination of a statistically relevant number of samples for the given period. As this is not possible at this point, the uncertainty of the sample-based parameters MUR, LR, HR and E_{calc} is therefore conservatively estimated at a maximum of 5%.

3.3 Fuels

The fuel type and preparation used in the calciner reactor influence the sorbent circulating between the reactors. Inactive (ash) and deactivated ($CaSO_4$) material is not able to absorb CO_2. Consequently, the input of impurities to the process by sulphur and ash cause performance losses by heating up and cooling down these circulating fractions.

To investigate the effect of fuel on the sorbent, long-term pilot operation was carried out with different types of fuels and varying particle sizes. Table 3.6 presents an overview of the coals used for flue gas production in the furnace as well as for providing the required heat for sorbent regeneration in the calciner reactor.

Table 3.6: Composition of fuels (as received).

Proximate analysis		HC 1	HC 2	HC 3	LEG/LEP
Fixed carbon	g/MJ	19.85	16.41	17.65	17.71
Water	g/MJ	4.73	3.44	1.90	4.32
Ash	g/MJ	5.87	2.92	2.35	1.66
Volatiles	g/MJ	10.57	12.63	11.61	19.94
LHV	MJ/kg	24.32	29.07	29.85	22.92
Ultimate analysis		HC 1	HC 2	HC 3	LEG/LEP
C	g/MJ	26.15	24.91	24.49	26.13
H	g/MJ	1.50	1.74	1.53	2.08
S	g/MJ	0.26	0.29	0.19	0.14
N	g/MJ	0.64	0.53	0.29	0.30
O	g/MJ	1.96	1.57	2.76	9.00
H2O	g/MJ	4.73	2.44	1.90	4.32
Ash	g/MJ	5.87	2.92	2.49	1.66
d_{50}	μm	45/1,500	45	45/500	35/500

The composition of the used coal types varies in sulphur and ash content. The amount of each species differs in relation to the LHV. Therefore, three different Hard Coals (HC) with sulphur contents of

0.19-0.26 g/MJ and ash contents of 2.49-5.87 g/MJ were used as calciner fuel. The pre-dried Lignite Energy Pulverized/Grained (LEG/LEP) contains less ash and sulphur. The sulphur and ash contents are 0.14 g/MJ and 1.66 g/MJ, respectively.

The hard coal (HC1 and HC2) was used either in coarse (d_{50} = 1,500 µm, sieved up to 10 mm) or in pulverized (d_{50} = 45 µm) particle size in the calciner. In addition, another hard coal (HC3) was fired with a d_{50} of 45 µm and 500 µm. Pre-dried Rhenish lignite in pulverized (LEG, d_{50}=36 µm) and coarse (LEP, d_{50}=500 µm) form was fuelled as well.

3.4 Sorbent

The sorbent utilized for the CaL process is natural limestone, a sedimentary rock that consists mainly of calcium carbonate. It is naturally available and is gained at low costs by open-cast mining worldwide. Natural limestone can be further utilized as raw material, e. g. in the cement industry.

The limestone used for the investigation of the CaL process was supplied by Rheinkalk GmbH, a subsidiary of the Lhoist group, the world's largest producer of lime and dolomite products. Since limestone is a natural product, the composition, porosity and other properties are not constant for each type of limestone. The CaL process demands a sorbent with a low cyclic deactivation combined with a large surface area and mechanical stability at the same time.

The sorbent is exposed to mechanical and thermal stress during operation causing attrition and sintering. This leads to a shift of the particle size distribution towards smaller fractions. Thus, Lhoist has developed a process to investigate the hardness of the sorbent. Sorbents with a hardness >5 J/m^2 are feasible for applications in fluidized bed systems [133]. The Lhoist limestone Messinghausen Fine (MF) has a sufficient hardness as oxide and carbonate. Thus, it was decided to utilize this limestone for pilot testing. Limestone gained from a different origin, Istein Fine (IF), with comparable mechanical properties was chosen to obtain complementary results.

The chemical composition of both types of limestone is shown in Table 3.7. The analysis show that they both consist of almost pure $CaCO_3$, 98.2 and 97.0 wt. respectively, with some minor inert fractions.

Table 3.7: Chemical composition of utilized limestone in wt.%.

	$CaCO_3$	$MgCO_3$	SiO_2	H_2O	Fe_2O_3	Al_2O_3	Other
MF 100-300 µm [134]	98.2	1.2	0.4	<0.1	<0.1	0.1	<0.2
IF 100-300 µm [135]	97.0	1.0	0.0	<0.1	0.3	0.4	<0.1

Both chosen types of limestone are also commercially available and have the required grain size for the application in the 1 MW$_{th}$ pilot plant to guarantee the required fluidization properties during operation. The particle size distribution of both types of limestone is shown in Fig. 3.4. The median diameters

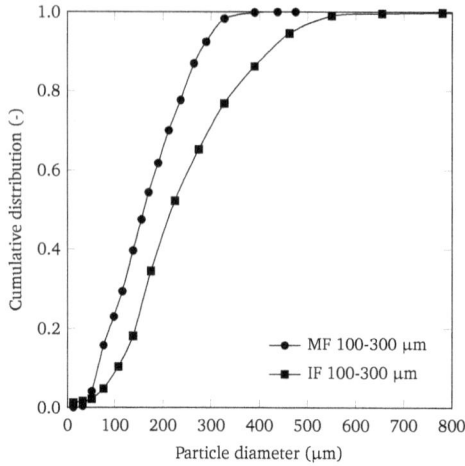

Figure 3.4: Particle size distribution of the utilized limestone.

d_{50} are 180 μm for MF and 220 μm for IF, respectively. Since the fluidization of the carbonator is slower (2.5-3.0 m/s) than the calciner (5-6 m/s), the lower fluidization velocities set the relevant range for the entrainment of the solids. At these velocities, particles < 500 μm are sufficiently fluidized. As depicted in Fig. 3.4, both types of limestone have considerably small fractions > 500 μm. So, all particles are sufficiently fluidized.

4 Results of Experiments

Between September 2015 and December 2017, the author was responsible for the preparation, conduction and evaluation of long-term pilot tests of the CaL process at the Institute of Energy Systems and Technology of TU Darmstadt. Six long -term test campaigns of four weeks each had the objective to gain the necessary data for scaling-up the technology to commercial size. The data also enabled the validation of process and CFD modelling. The experimental operation was as close as possible to industrial conditions such as oxy-fuel calcination and coal-originated flue gas for decarbonization. The first four test campaigns were conducted between September 2015 and April 2016, followed by another two test campaigns from September to December 2017. All in all, the plant was operated $> 3,000$ hours, thereof 2,000 hours in coupled CFB mode capturing CO_2 for more than 1,500 hours.

The first period of test campaigns was used to accomplish steady-state long-term operation under a broad range of operating conditions. Different types of fuels with different particle size distributions were fuelled in the calciner. Also, different types of limestone were used. The second period of pilot testing was based on the findings of the previous four test campaigns. Optimized operation was identified and a new coupling concept from calciner to carbonator with a cone valve was successfully tested with regard to later application in a demonstration plant.

This chapter describes the range of parameters during operation and shows the approach of steady-state operation. Additionally, the temperature and pressure profiles of both carbonator and calciner reactor as well as the hydrodynamics of the utilized coupling concepts between the reactors are shown. The main operating parameters for carbonator and calciner are assessed to identify the basis for the scale-up to an demonstration plant. The results have already been published [29, 125, 126, 136].

4.1 Steady-State Operation

The important objective of the research was the achievement of stationary operating points, so-called steady-state operation. Steady-state operation comprises the stable conditions in sorbent and especially in the solid sorbent phases. Particular focus during operation was set on achieving these representative operational conditions. In particular, the steady-state conditions in the sorbent phase are of great interest since reliable results, e. g. sorbent conversion or accumulation of impurities, can only be attained this way. Due to the slow sorbent deactivation, assuring steady-state operation in the sorbent phase required long operational periods with stationary operation. Depending on the continuous feed of make-up to influence the residual activity of the sorbent, a homogeneous mixing of the inventory takes several ten hours up to days.

The challenge of achieving steady-state operation is to determine a sufficient period of constant operation without significant changes of operational variables. Steady-state operation in the gas phase can be determined comparatively easily. The recorded measurements data of reactor pressures, temperatures, make-up and coal mass flows as well as flow and gas concentrations of all in- and outgoing gas streams deliver a good picture during operation. In contrast, the conditions in the sorbent phase are examined by samples analysis weeks after operation. Thus, especially in the pilot test carried out first, it was important to ensure sufficient periods of stable operation. Therefore, the mean residence time of the sorbent was taken into account to estimate the required period to reach steady-state conditions in the solid phase. Three times the residence time of the sorbent in the system was utilized to ensure that at least 95 % of the inventory was exchanged by the make-up flow. Up to 60 hours of stable operation were conducted in the first test runs to ensure steady-state operation with high certainty.

To define steady-state operating points for evaluation, approx. 200 recorded measurements of temperature, mass and volumetric flow rate, level and gas concentration had to be considered. For each value, the coefficient of variation was calculated as the normalized variance to describe the relative scattering value. An hourly coefficient of variation for all values was used as an criterion for the stability of the operating period. On the basis of earlier investigations and evaluations of measurement uncertainties, a stability criterion of $\leq 5\%$ was applied. A final validation of steady-state operation was then given by the analyzed solid samples for the chosen operating points. A detailed description of the steady-state consideration for all test runs was published by Helbig [125, 132].

Besides the steady-state operation, some transient effects and influences were evaluated. In order to determine transient effects, a minimum degree of stability of the operating point is required. For example, there are effects that are particularly well shown in transient processes since these usually run on very small time scales. These procedural parameters are subject to only minor changes during short periods, e.g. the carbonator temperature to change the maximum absorption efficiency limited by the thermodynamic equilibrium. Changing one of these parameters after achieving a steady-state operating point allows the observation of transient effects.

As described above steady-state long-term operation is crucial. It is important to describe the approach to achieve steady-state operating periods. Fig. 4.1 shows an exemplary period of pilot tests. In general, an experimental test run started with the heating of the dual fluidized bed system following the temperature gradients of the refractory lining. In the beginning, both reactors were fluidized with electrically pre-heated air followed by the start-up burners. Sorbent was filled into the reactor system, and pressure sealing was achieved in the loop seals (see Fig. 4.1-a). Afterwards, solid circulation was started and the mass flow from the carbonator to the calciner was stepwise increased between 0 to 5 h, as shown in Fig. 4.1-c. Stable hydrodynamics in the dual CFB system with a minimum of circulating sorbent was established. The circulating material heated up all interconnecting parts of the system. The reactors were continuously heated, and after reaching 700 °C in the calciner at 3 h, coal firing was started to substitute the gas burner and was increased stepwise (see Fig. 4.1-d). The calcination of the sorbent took place as the calciner was heated up to operating temperature. The solid circulation was further increased at 5 h and utilized as a heat carrier to the carbonator to stop the start-up burner. At 4 h, the carbonator fluidization air was then enriched with CO_2 to start the CaL process (see Fig. 4.1-b). The

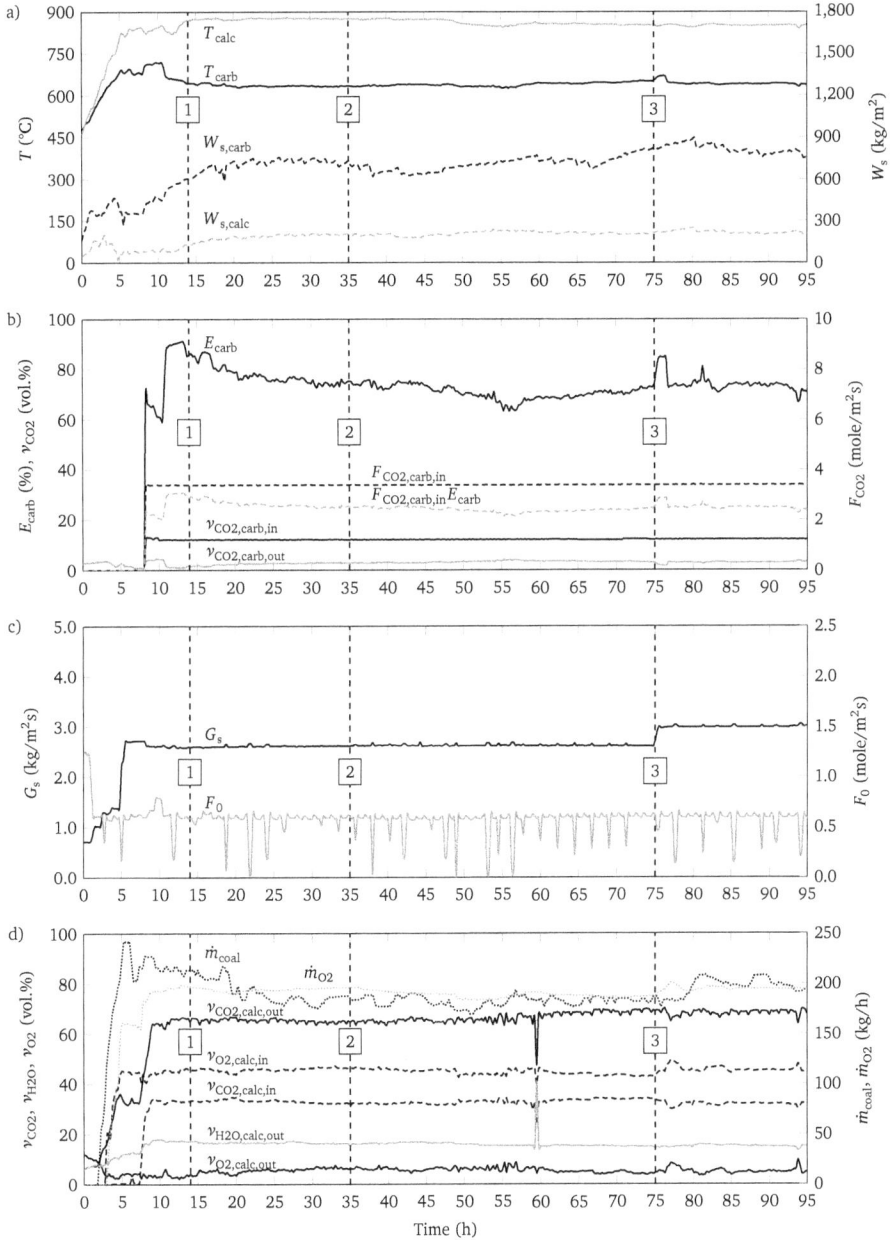

Figure 4.1: Long-term steady-state operation .

starting carbonation reaction increased the reactor temperature that was controlled by the bayonet cooling tubes (see Fig. 4.1-a). The solid circulation was adapted gradually to achieve stable CO_2 absorption in the carbonator. After establishing stable operation of the total system, the calciner operation was switched from air- to oxy-combustion mode by replacing air with oxygen enriched flue gas at 7 h (see Fig. 4.1-d). The carbonator fluidization was switched from artificial to coal originated flue gas from the combustion chamber at 11 h. A continuous feed of make-up was established to adjust the sorbent activity as well as to replace the loss of fines due to attrition (see Fig. 4.1-c). Material was extracted from the reactors to keep the inventory constant while feeding make-up. At 14 h (1), all parameters were set for the continuous and stable operation.

Between 14 h (1) and 35 h (2), the process ran into steady-state conditions. The temperatures T_{calc} of 880 °C in the calciner and T_{carb} of 640 °C in the carbonator were kept constant. The specific inventory of the carbonator $W_{s,carb}$ was adjusted to 700 kg/m^2 and the calciner $W_{s,calc}$ to 200 kg/m^2, respectively (see Fig. 4.1-a). The solid circulation rate between the reactors G_s was set to 2.6 kg/m^{2}s corresponding to a molar sorbent looping ratio of 13 mole$_{Ca}$/mole$_{CO2}$. 0.6 mole/m^{2}s of make-up were fed corresponding to a make up ratio of 0.17 mole$_{Ca}$/mole$_{CO2}$ (see Fig. 4.1-c). The calciner ran in oxy-fuel conditions, i. e. flue gas was recirculated and enriched with oxygen to avoid dilution. The recirculated flue gas contains 34 vol.% of CO_2 and 46 vol.% of O_2 balanced with some residual N_2 and H_2O. At the beginning of the period, 209 kg/h of coal and 197 kg/h of O_2 were fed to keep the calciner temperature constant. The calciner off gas contained 66 vol.% of CO_2, 17 vol.% of H_2O and 3 vol.% of O_2 (see Fig. 4.1-d). The process continuously ran for 21 hours without significant changes of the control parameters. Fig. 4.1-b shows the development of the carbonator absorption efficiency E_{carb} over this period. While a constant flow $F_{CO2,carb,in}$ of 3.4 mole/m^{2}s was fed with a volumetric CO_2 concentration $v_{CO2,carb,in}$ of 12.2 vol.%, the amount of CO_2 absorbed decreased during the period. At 14 h, 3 mole/m^{2}s were captured corresponding to an E_{carb} of 88 %. As can be seen, the carbonator efficiency decreased to a residual value of 78 % and 2.5 mole/m^{2}s of CO_2 were captured. The process ran into steady-state conditions during the period as the CO_2 carrying capacity decreased by sorbent deactivation. During this period, the calciner fuel was slightly decreased from 209 to 196 kg/h. Since all the sorbent had undergone first calcination and less CO_2 had to be released from the circulating sorbent, the fuel input could be lowered. The oxygen input was kept constant to operate the calciner with an increased oxygen excess to improve the combustion efficiency of the coal (see Fig. 4.1-d).

After reaching steady-state conditions at 35 h (2), the actual setup was used to investigate the influence of the reactor temperatures. Therefore, either the calciner temperature was lowered or the carbonator temperature was increased during this phase. The changes were made stepwise (see Fig. 4.1-a). At 55 h, the effect of the decreased calciner temperature to 850 °C is shown. Lowering the calciner temperature led to an insufficient calcination efficiency E_{calc} and the sorbent was not fully regenerated. As a consequence, the circulating solid stream can absorb less CO_2 in the carbonator which leads to a decrease of the carbonator efficiency E_{carb} down to 64 %. At 75 h (3) a major change was made and the solid circulation between the reactors was increased from 2.6 to 3.1 kg/m^{2}s increasing sorbent looping ratio from 13 to 15 $\text{mole}_{Ca}/\text{mole}_{CO2}$. As can be seen in Fig. 4.1-b, the carbonator efficiency is not significantly influenced. At the same time, the oxygen and coal mass flows were increased to keep the calciner temperatures stable (see Fig. 4.1-d). After 20 hours of continuous operation, another steady-state was achieved. The approach described in the previous paragraphs was used to conduct long-term pilot tests to gain reliable data for the scale-up of the technology.

4.2 Heat and Mass Balance

The mass balance is based on the fact that a system's total change of mass corresponds to the difference of in- and outgoing mass. For steady-state systems where changes of the mass do not occur, the entering and leaving mass flows are equal.

The heat flows are balanced on the basis of the first axiom of thermodynamics. The balance considers all forms of energy in a thermodynamic system. The total energy of a system changes by supply or removal of energy across the system boundaries by heat, work or energy transported by mass. The balancing of stationary flow processes is used for the evaluation since in- and outcoming mass flows are identical and the system energy as a function of time is constant. Furthermore, the kinetic and potential changes of the mass flows are neglected. Using the specific enthalpy h, Eq. 4.1 results in a simplified form of the first axiom of thermodynamics for energy balancing of stationary flow processes [137].

$$0 = \dot{Q} + P + \sum \dot{m}_{i,in} h_{i,in} - \sum \dot{m}_{i,out} h_{i,out} \qquad (4.1)$$

In Eq. 4.1, the heat sources and sinks are expressed by $\dot{Q}$, the power input by P and the flows of energy transported across system boundaries by $\dot{m}_i h_i$. These integral parts are derived for the balance sheets of carbonator and calciner as depicted in Fig. 4.2.

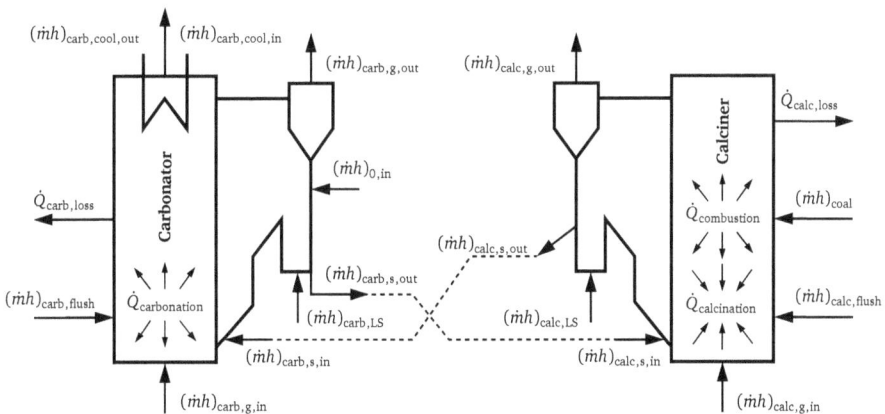

Figure 4.2: Balance sheet of carbonator and calciner for the directly heated CaL pilot plant.

The carbonator reactor system is balanced as a total system with in- and outgoing gas and solid streams of all components that are incorporated into the reactor and the peripheral components such as the loop seal. The balance considers the reaction heat of the carbonization reaction as a heat source and the heat dissipated by the cooling system as a heat sink. The extracted heat by the cooling lances is calculated with the mass flow of cooling water and the temperature difference of the inlet and outlet flow. The heat loss of the carbonator and the peripheral components are also considered. The heat loss is estimated by the reactor surface area and the measured surface temperature. The solid flow circulating between the reactors is calculated by the characteristic curve of the screw conveyor and the composition of the solid samples analyzed from the considered operating point.

The results of the carbonator heat and mass balance for an exemplary steady-state operating point is shown in Table 4.1. All entering streams are indicated with a plus, all leaving with a minus. The heat and mass balance closure shows a good agreement of in- and outgoing streams with a deviation for the mass balance of 0.5 % and for the heat balance of 0.2 %, respectively.

Table 4.1: Heat and mass balance of the carbonator.

Type		$\dot{m}$ (kg/h)	$\dot{Q}$ (kW)
Gas	$\sum$ inlet	+1132.7	+37.9
	$\sum$ outlet	-997.4	-176.2
Solids	$\sum$ inlet	+4156.7	+873.5
	$\sum$ outlet	-4350.4	-610.2
Sources & sinks	Carbonation	-	+115.9
	Cooling	-	-230.8
	Losses	-	-28.1
Balance	$\sum$ inlet	+5332.5	+1019.1
	$\sum$ outlet	-5304.6	-1017.6
	Δ absolute	28.1	1.9
	Δ relative	0.5 %	0.2 %

The calciner reactor system balance includes all in- and outgoing gas and solid streams that are required to run the process. As already mentioned at the carbonator balance, the internal solid recirculation is not considered. The calcination reaction of the carbonate content of the circulating solid stream constitutes a heat sink. The combustion of the coal is depicted as a heat source. The released heat of the combustion process is derived by a combustion calculation. It is assumed that volatiles are completely converted. Based on sample analysis and mass balancing, approx. 10 % of the carbon content is not converted and leaves the system with the fly ash that is considered in the combustion calculation. The heat losses estimated by the reactor surface and temperature are considered as another heat sink.

Table 4.2 shows the complementary heat and mass balance of the calciner for the same operating point described for the carbonator before. The mass balance closure shows a deviation of 0.3 % and the heat balance of 1.0 %, respectively. The higher deviations can be explained by the uncertainties regarding the burnout of the coal and thus the released heat in the reactor.

Table 4.2: Heat and mass balance of the calciner.

Type		$\dot{m}$ (kg/h)	$\dot{Q}$ (kW)
Gas	$\sum$ inlet	+586.4	+60.9
	$\sum$ outlet	-4435.3	-250.1
Solids	$\sum$ inlet	+805.8	+574.1
	$\sum$ outlet	-4204.9	-883.2
Sources & sinks	Calcination	-	-124.7
	Combustion	-	+663.4
	Losses	-	-51.9
Balance	$\sum$ inlet	+4890.8	+1298.3
	$\sum$ outlet	-5021.7	-1309.9
	Δ absolute	10.9	11.7
	Δ relative	0.3 %	1.0 %

4.3 Operational Range

The performance of the CaL process depends on a set of specific process-engineering parameters. The range of operational and evaluation parameters of the process, for example CO_2 capture efficiency or molar Ca sorbent conversion, is shown in Table 4.3. A large number of the process parameters cannot be determined directly from the test results and is based on a detailed knowledge of the mass and energy flows of the pilot plant obtained by balancing the process.

Table 4.3: Range of operating parameters during long-term pilot tests.

Description	Parameter	Unit	Range
Carbonator average temperature	T_{carb}	°C	620-695
Carbonator inventory	$W_{s,carb}$	kg/m^2	330-920
Carbonator superficial gas velocity	$u_{0,carb}$	m/s	2.0-3.0
Carbonator molar CO_2 inlet flow	$F_{CO2,carb,in}$	$mole/m^2 s$	2.8-5.5
Carbonator volumetric CO_2 inlet fraction	$v_{CO2,carb,in}$	vol.%	10.4-15.5
Carbonator efficiency	E_{carb}	%	44-95
Carbonator molar conversion	X_{carb}	$mole_{CO2}/mole_{Ca}$	0.02-0.11

Table 4.3: Range of operating parameters during long-term pilot tests.

Description	Parameter	Unit	Range
Solid circulation	G_s	kg/m²s	1.9-4.2
Sorbent looping ratio	LR	$mole_{Ca}/mole_{CO2}$	6-20
Make-up flow	F_0	mole/m²s	0.1-1.0
Make-up ratio	MUR	$mole_{Ca}/mole_{CO2}$	0.05-0.24
Calciner average temperature	T_{calc}	°C	820-920
Calciner inventory	$W_{s,calc}$	kg/m²	60-350
Calciner superficial gas velocity	$u_{0,calc}$	m/s	4.5-6.0
Calciner heat input	$\dot{Q}_{th,calc}$	kW$_{th}$	625-1,310
Calciner molar conversion	X_{calc}	$mole_{CO2}/mole_{Ca}$	0-0.06
Calciner volumetric CO_2 inlet fraction	$v_{CO2,calc,in}$	vol.%	0-40
Calciner volumetric O_2 inlet fraction	$v_{O2,calc,in}$	vol.%	29-60
Calciner volumetric H_2O inlet fraction	$v_{H2O,calc,in}$	vol.%	0-26
Calciner volumetric CO_2 outlet fraction	$v_{CO2,calc,out}$	vol.%	36-80
Calciner volumetric O_2 outlet fraction	$v_{O2,calc,out}$	vol.%	0.6-8.2
Calciner volumetric H_2O outlet fraction	$v_{H2O,calc,out}$	vol.%	11-32

4.3.1 Hydrodynamics

A main challenge in the testing and scale-up process is the hydrodynamic behaviour of the CaL plant. The required process parameters can be met by a stable pressure profile in the reactors and its interconnections in order to provide a well-designed sorbent distribution and a continuous sorbent flow to achieve steady-state operating conditions. Therefore, pilot tests with different coupling concepts with focus on coupling calciner and carbonator reactor were carried out.

Figs. 4.3 and 4.4 show the characteristic pressure profiles of two different dual CFB configurations. Both configurations were operated with an identical carbonator setup, i. e. the solids were transferred from carbonator to calciner with a screw conveyor. The coupling concept to transfer the solids from calciner to carbonator was varied (see Sec. 3.1, Fig. 3.2). Configuration I coupled the calciner to the carbonator with two loop seals (Fig. 4.3-b), in Configuration II a loop seal equipped with a cone valve was used (Fig. 4.4-b). The carbonator pressure profiles for both configurations show a sharp increase in the lower region which indicates that the main share of the inventory is in the dense region in the lower part of the reactor (see Figs. 4.3-a and 4.4-a). This solid distribution is caused by fluidization velocity of 2.5-3.0 m/s corresponding to the lower range of the fast fluidization pattern with good gas-solid contact.

In Configuration I, the calciner was operated without internal recirculation, and two loop seals were used to transfer the solids to the carbonator (see Fig. 4.3-b). The calciner pressure profile is smoothly shaped in the bottom region with an almost linear gradient in the riser region indicating a uniform distribution of the inventory along the reactor height. In this configuration, the solid particles leave the calciner reactor without being recirculated. This coupling concept offers very stable hydrodynamic conditions since the calciner inventory can be kept constant by controlling the reactor velocity between 4.5-5.5 m/s at certain sorbent looping ratio. Nevertheless, a degree of freedom is removed from the system by abandoning the internal solid circulation of the reactor. The operational long-term experience with high sorbent regeneration efficiencies confirmed the robustness of this concept, as shown in recent publications [29, 125, 126].

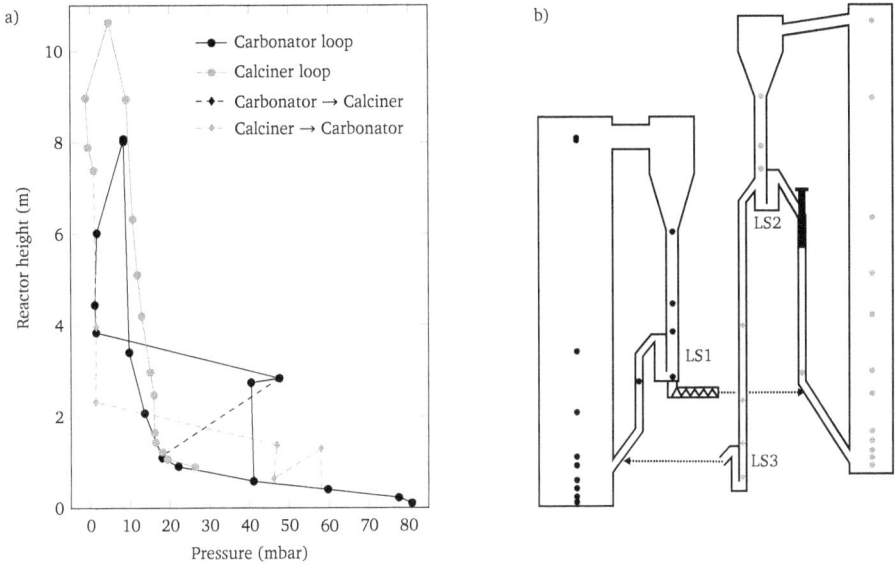

Figure 4.3: Configuration I: Pressure profiles of both reactors with closed pressure cycle with circulation path (a) and reactor setup for calciner to carbonator coupling with two loop seals (b).

Configuration II (see Fig. 4.4-b) offers different hydrodynamic and operational aspects. The calciner is operated with internal recirculation, and the pressure profile is similar to Configuration I. The lower loop seal is removed and a cone valve is installed in the upper loop seal to control the solid flow. Thus, the superficial gas velocity in the calciner could be decoupled from the external solid circulation between the reactors. The solid inventory in the calciner can be controlled by the opening of the cone valve while the fluidization is kept constant.

Pilot operation with Configuration II is shown in Fig. 4.5. The pilot plant was operated under steady-state conditions in the initial period for 7 h. The average temperatures in the carbonator and the calciner were constant at 650 °C and 880 °C, respectively. In this phase, the absorption efficiency in the carbonator was around 72 % (see Fig. 4.5-a). The carbonator fluidization with coal-originated flue gas from the

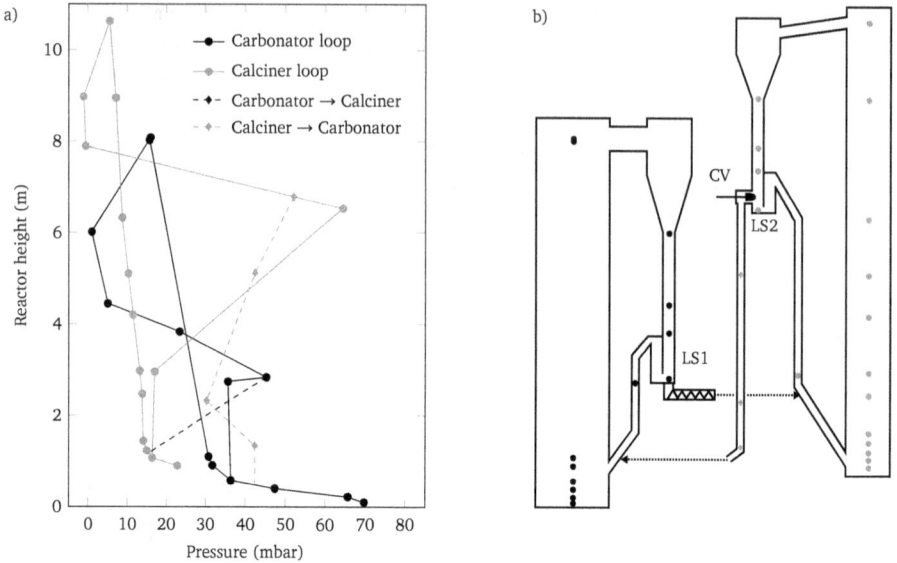

Figure 4.4: Configuration II:Pressure profiles of both reactors with closed pressure cycle with circulation path (a) and reactor setup for loop seal-cone valve calciner to carbonator coupling (b).

combustion chamber led to a superficial gas velocity of 2.2 m/s. The specific reactor inventory was kept at around 650 kg/m². The calciner was operated under oxy-fuel combustion fluidized with a mixture of oxygen and recirculated flue gas with a superficial gas velocity of 5.0 m/s. The inventory was kept at around 150 kg/m² (see Fig. 4.5-b). During the period, make-up was constantly fed with a feeding rate of 0.58 mole/m²s and mixed with the inventory and the circulating solid stream of 2.6 kg/m²s. The solid circulation between the reactors was kept constant by the screw conveyor. The cone valve opening position X_{cv} was kept between 52 and 55 % (see Fig. 4.5-c). Small movements of the cone valve, as can be seen at 3 h, were required to avoid blocking with respect to solid particles build up in and around the cone valve opening.

At 7 h (1), the solid circulation stream from carbonator to calciner was increased from 2.6 to 3.0 kg/m²s by changing the rotation speed of the screw conveyor. At the same time, the cone valve position was changed to balance the solid flow vice versa. Therefore, the cone valve was opened from 55 to 70 % and then closed stepwise to 60 % to keep the inventory in the calciner and carbonator system constant, as can be seen in Fig. 4.5-b. To maintain the calciner reactor temperature, the thermal power in the calciner had to be increased. The coal and the oxygen feeding rate as well as the recirculation gas was slightly increased. As a consequence, the superficial gas velocity in the calciner increased to 5.5 m/s, while the superficial gas velocity in the carbonator was kept constant (see Fig. 4.5-b). As a result of increased solid circulation between the reactors at 10 h (2), the carbonator efficiency rose to 82 % (see Fig. 4.5-a).

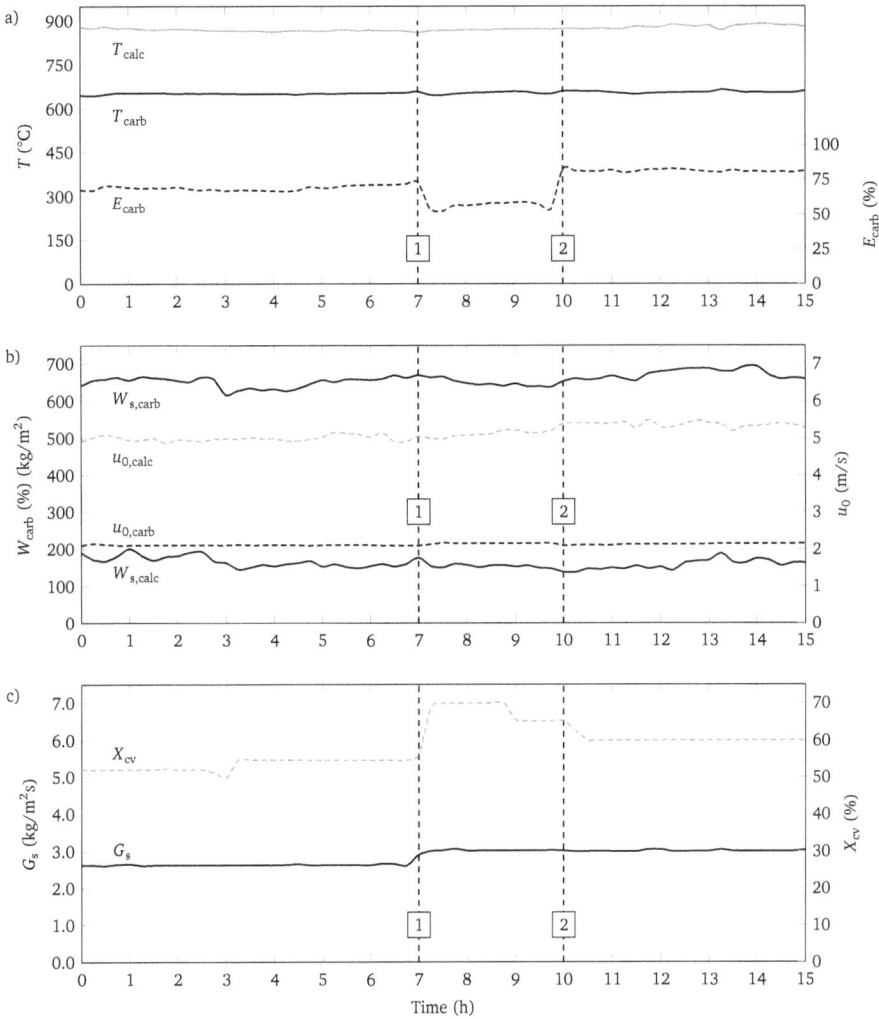

Figure 4.5: Exemplary operation of calciner to carbonator coupling with a loop seal and cone valve (Configuration II).

The exemplary operation with the advanced coupling concept utilizing a loop seal equipped with a cone valve to transfer the circulating solid from calciner to carbonator showed stable pilot operation and good controllability for the hydrodynamics of the plant. The inventory of the calciner could be controlled by changing the position of the cone valve while increasing the superficial gas velocity in the calciner reactor.

The reactor temperatures of carbonator and calciner are important for the absorption of CO_2 and the regeneration of the sorbent. Exemplary temperature profiles both of carbonator and calciner are depicted in Fig. 4.6 showing various operating points.

In the carbonator (see Fig. 4.6-a), all temperature profiles show decreasing temperatures with increasing reactor height. The temperature profile is shaped by the bayonet cooling since cooling lances are lowered from the top into the reactor. Three to five lances can be axially arranged down to a level of 3 meters above the nozzle grid. The less cooling lances are used, the more homogeneous is the temperature profile. The particles and the gas are cooled down before entering the cyclone. Most of the entrained particles are recirculated to the bottom zone where the exothermic carbonation reaction runs. The internally recirculated particles cool down the reactor to adjust the reactor temperature. Thus, a warmer dense zone and a colder lean zone is found. An average carbonator temperature is determined from the reactor profile and applied for the evaluation. In-bed gas analysis taken at various operating points show that the main part of the CO_2 is absorbed in the dense zone of the reactor (approx. 85 %). Equilibrium conditions could be achieved in the dense zone. The carbonation reaction runs faster since the higher dense zone temperatures positively affect the reaction kinetics. The lower temperatures in the lean zone

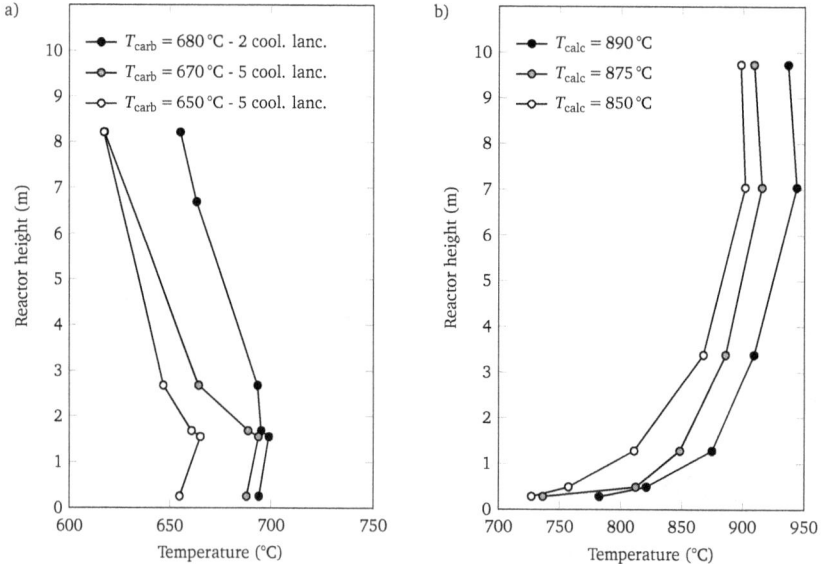

Figure 4.6: Exemplary carbonator (a) and calciner (b) temperature profiles along the reactor height for different operating temperatures.

lead to a lower equilibrium. Thus, carbonation reaction can continue there in order to achieve high carbonator absorption efficiencies.

In comparison with the carbonator, the calciner temperatures (see Fig. 4.6-b) show an inverted profile with increasing reactor height. At the bottom of the reactor, the sorbent from the carbonator, the coal and the fluidization gas enter the system and need to be heated up, and the endothermic calcination reaction consumes heat. In addition, the reactor bottom of the the calciner is formed as a cone with increasing diameter from 280 mm to 400 mm. As a consequence, the particles have only a short residence time due to locally increased gas velocities in the cone. Further up in the reactor, calcination and combustion take place. A stable temperature close to or above 900 °C is reached that allows highly efficient sorbent regeneration.

4.4 Carbonator Evaluation

The efficiency of the carbonator is essential for the performance of the CaL technology. The absorption efficiency is determined by a variety of parameters, such as reactor temperature, the solids inventory or the sorbent properties. Many of these parameters correlate with each other, and thus considerably complicate the evaluation of isolated effects. This section discusses the different impact factors on the carbonator performance.

4.4.1 Temperature

Since the carbonator efficiency as a crucial parameter is strongly dependent on the reactor temperature, the results from pilot operation were evaluated with a focus on this relation. For scale-up purposes, a matter of particular interest is the definition of an operating window for the carbonator temperature. On the one hand, the carbonator temperature should be low to decrease the equilibrium CO_2 concentration, and thus to enable a high capture rate. On the other hand, the carbonator temperature should be as high as possible to avoid inhibition of the reaction kinetics and to minimize the fuel input required in the calciner to heat up the cold solid stream entering from the carbonator.

Exemplary steady-state carbonator operation is presented in Fig. 4.7 to show the direct influence of the reactor temperature. The system was steadily operated with constant inventory, solid circulation, make-up feed under oxy-calcination conditions. Thereby, the carbonator was fluidized with a flue gas stream containing a CO_2 inlet concentration $y_{CO2,carb,in}$ of 13 vol.%. The carbonator temperature T_{carb} was kept constant at around 675 °C in the initial period for 20 h (1). During this period, the CO_2 outlet concentration could be reduced significantly to 2.7 vol.%, resulting in a CO_2 absorption efficiency of 80 % in the carbonator. The maximum absorption efficiency limited by the equilibrium ($E_{carb,eq}$) was 86 %. At 20 h, the gas pre-heating was lowered and thus, T_{carb} was slightly raised to 685 °C. The direct consequence can be seen in the absorption efficiency that dropped to 75 % which can be explained by the increasing equilibrium CO_2 partial pressure. The maximum absorption at this temperature level was 81 %. In the next step after 23 h (2), the heat extraction by the cooling lances was increased to lower T_{carb} to 650 °C. A significant increase of E_{carb} to 88 % can be observed, and operation close to $E_{carb,eq}$ of 92 % was achieved. The tests show the crucial influence of the carbonator temperature on the absorption efficiency. Thereby, the maximum carbonator absorption efficiency $E_{carb,eq}$ limiting carbonator efficiency by the thermodynamic equilibrium as a function of the reactor temperature could be reached by 95 %. Opposing effects in the short-term such as a decrease of E_{carb} while T_{carb} decreases at the same time, as shown at 10 h, can be explained by fluctuations in circulating sorbent stream coming from the calciner.

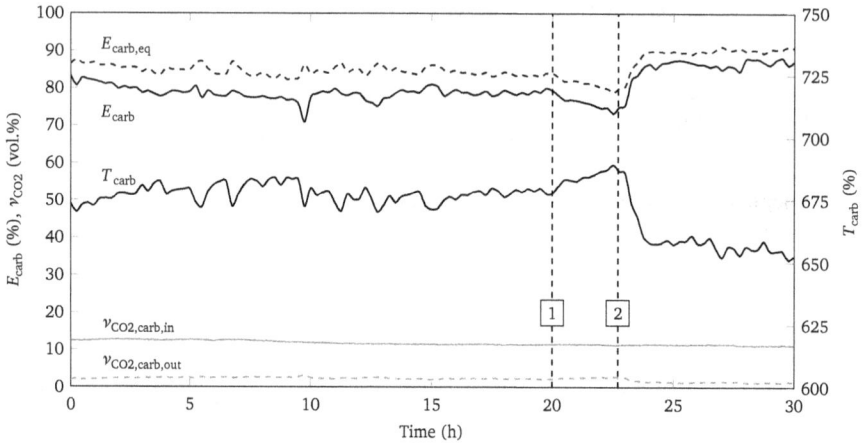

Figure 4.7: Steady-state carbonator operation at different reactor temperatures.

The results of pilot operation with high capture rates close to the equilibrium conditions in the range between 625 °C and 700 °C of the described period above is shown in Fig. 4.8. The measured temperatures are plotted to screen out the effect of the carbonator temperature under steady-state conditions with a highly active sorbent that allows operation close to the thermodynamic equilibrium. In order to achieve carbonator absorption rates >80 %, the best suitable operational range of 650 °C to 675 °C is derived from the pilot tests. In this operational windows, carbonator absorption rates in a range of 80 to 90 % is possible. Lower temperatures (<650 °C) are also theoretically feasible to achieve carbonator absorption rates up to 90 %, but a tendency to lower efficiency with decreasing temperature is shown. Lower temperatures slow down the carbonation reaction. Due to the reduced reaction rate, the particle residence time is too short and the gas-solid contact too low to realize a complete conversion. The increasing demand of fuel input to the calciner to heat up the entering solids for calcination is an additional disadvantage of lower carbonator temperatures. Higher temperatures (>675 °C) are not feasible for highly efficient capture since the equilibrium CO_2 concentration strongly increases with increasing reactor temperature. As indicated by the trend in Fig. 4.8, the maximum absorption efficiency in the carbonator is limited to <80 % at T_{carb}>675 °C.

4 Results of Experiments

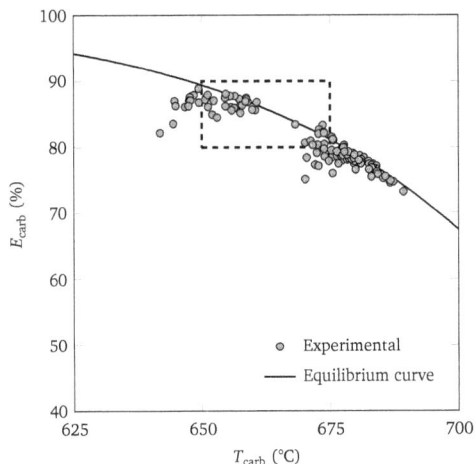

Figure 4.8: Carbonator absorption efficiency over carbonator temperature for the steady-state operation period shown in Fig. 4.7.

4.4.2 Closure of CO_2 Balance

An important factor of process evaluation is the closure of the CO_2 mass balance of the carbonator. During steady-state operation, the CO_2 removed from the gas phase as well as inserted CO_2 with the make-up can be calculated and compared with the $CaCO_3$ formed in the circulating stream of CaO. Hence, the following mass balance has to be fulfilled:

$$F_{CO2,carb,in}E_{carb} = F_{Ca}(X_{carb} - X_{calc})$$ (4.2)

The CO_2 removed from the gas phase can be calculated by the flow and gas composition measurement at the inlet and outlet of the carbonator. Thus, the terms on the left side of Eq. 4.2 are most reliable. The solid phase on the right side of Eq. 4.2 includes the molar flow F_{Ca} representing the amount of CaO particles that enter the carbonator with a molar carbonate content X_{calc} after regeneration and leave the reactor after absorption with the carbonate content X_{carb}. F_{Ca} is calculated from the total circulating solid mass flow G_s and the solid samples taken from the reactor system, as described by Eq. 3.1 in Sec. 3.2.

The two terms of Eq. 4.2 are compared in Fig. 4.9. The CO_2 removed from the gas phase in the carbonator and the increase of $CaCO_3$ in solid circulating flow between the reactors is depicted. The comparison includes the results from various operating points with sample analysis available, operated under a comprehensive range of parameters, e. g. reactor inventory and temperature, make-up feed, solid circulation. For all operating points, solid samples from the reactors were extracted and considered for each point in the balance. The CO_2 bound in the solid phase matches the CO_2 removed from the gas phase with a reasonable deviation, except for operating with coarse coal in the calciner. There, additional effects such as oxidation of unburned char from the calciner exert an influence on the balance.

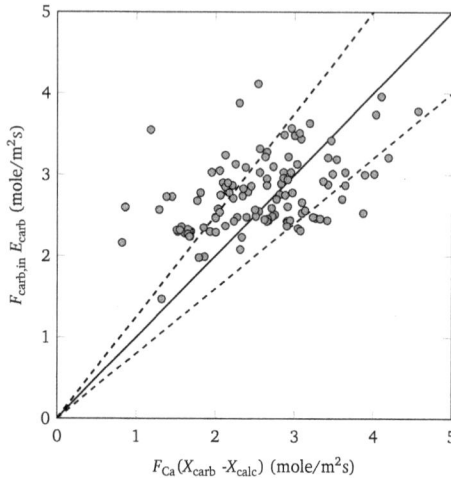

Figure 4.9: Comparison between removed CO_2 from the gas phase and the increment of $CaCO_3$ in the circulating solid stream between the reactors.

4.4.3 Active Space Time

The application of a basic reactor model to the experimental data is particularly useful to evaluate the results and to design the carbonator. Therefore, the model of the active space time combines the different sorbent and process parameters. This model is used in literature to compare the results of different pilot tests.

The model approach is based on the balance of CO_2 streams of the carbonator reactor as shown in Eq. 4.3. The balance of CO_2 absorbed in the gas phase (left term) and the amount of solids in the carbonator reacting with the CO_2 with an average reaction rate (right term) is used to build the model.

$$F_{CO2,carb,in} E_{carb} = n_{Ca,carb,active} \left(\frac{dX}{dt} \right)_{carb} \quad (4.3)$$

The reaction rate of the particles in the carbonator is based on the assumption that the present particles react with a constant rate up to a maximum conversion rate X_{avg}. A perfectly mixed reactor for the solid phase and a plug flow reactor for the gas phase are assumed. This model has been applied for various tests to evaluate experimental results [30, 105]. According to the methodology postulated by

4 Results of Experiments

Charitos et al. [105], the reaction rate depends on the sorbent constant k_s, the gas-solid contacting factor φ, the CO_2 carrying capacity X_{avg} and the difference between the average and equilibrium CO_2 volume fraction in the reactor. The reaction rate is shown in Eq. 4.4.

$$\left(\frac{dX}{dt}\right)_{carb} = k_s \varphi X_{avg} \left(\overline{\nu_{CO2,carb}} - \nu_{CO2,carb,eq}\right) \tag{4.4}$$

The active inventory $n_{Ca,carb,active}$ can now be defined based on the reaction rate term. According to the postulated model, only an active fraction ($f_{active,carb}$) of the particles $n_{Ca,carb}$ takes part in the kinetically-controlled fast reaction regime. The active fraction corresponds to particles with a residence time less than the time required to increase the carbonate content from X_{calc} to X_{avg} (t^*) according to Eq. 4.4.

$$n_{Ca,carb,active} = n_{Ca,carb} f_{active,carb} = n_{Ca,carb} \left(1 - e^{\frac{-t^*}{\bar{t}_{carb}}}\right) \tag{4.5}$$

The characteristic carbonation time t^* can be calculated with the results of solid sample analysis and the reaction rate model of Eq. 4.4.

$$t^* = \frac{X_{avg} - X_{calc}}{(dX/dt)_{carb}} = \frac{X_{avg} - X_{calc}}{k_s \varphi X_{avg} \left(\overline{\nu_{CO2,carb}} - \nu_{CO2,carb,eq}\right)} \tag{4.6}$$

The combination of Eqs. 4.3 to 4.6 leads to a model describing the carbonator absorption efficiency of a CaL system by including all directly or indirectly linked operating parameters in Eq. 4.7. The apparent reaction constant $k_s \varphi$ is thereby calculated by the comparison of CO_2 absorbed in the gas phase, and CO_2 reacted in the solid phase. The reaction rate constant k_s can be determined by thermogravimetric analysis of the sorbent. An average value of 0.45 1/s for different limestones is found in literature [102].

$$E_{carb} = \frac{n_{Ca,carb}}{F_{CO2,carb,in}} k_s \varphi f_{active,carb} X_{avg} \left(\overline{\nu_{CO2,carb}} - \nu_{CO2,carb,eq}\right) \tag{4.7}$$

For given operating conditions, the so-called active space time $\tau_{active,carb}$ (see Eq. 4.8) indicates the performance of the carbonator as a single parameter in Eq. 4.7. The inventory, the residence time of particles, the CO_2 concentrations and temperature as well as influences depending on the type of limestone are considered.

$$\tau_{active,carb} = \frac{n_{Ca,carb}}{F_{CO2,carb,in}} f_{active,carb} X_{avg} \tag{4.8}$$

The comparison of CO_2 removed from the gas phase and the CO_2 reaction in the carbonator bed applying the active space time is shown in Fig. 4.10. A reasonable agreement of the model based on average testing conditions ($\nu_{CO2,carb,eq} = 1.1$ vol.%, $\nu_{CO2,carb,in} = 13$ vol.%, $k_s \varphi = 0.31$ 1/s) with the experimental data is accomplished. The apparent carbonation rate constant within the carbonator reactor ($k_s \varphi$) was derived from TGA experiments by Helbig [132]. An averaged value of 0.45 1/s was obtained with a contacting efficiency φ of 0.7. A simplification was set to apply the active space time model to the experimental data. As reported by Charitos et al. before [105], the values of X_{avg} and X_{carb} are very close for residence times around 3 minutes in the carbonator. This was proven in TGA experiments by Reitz [138] for low carbonate contents $X_{carb} < 0.1$ mole$_{CaCO3}$/mole$_{Ca}$. In pilot operation, the carbonator was operated with a residence time of particles of 3-8 minutes with X_{carb} of 0.02-0.1 mole$_{CaCO3}$/mole$_{Ca}$ and consequently, the simplification was made to use X_{carb} instead of X_{avg} for the evaluation of the experimental data in this chapter.

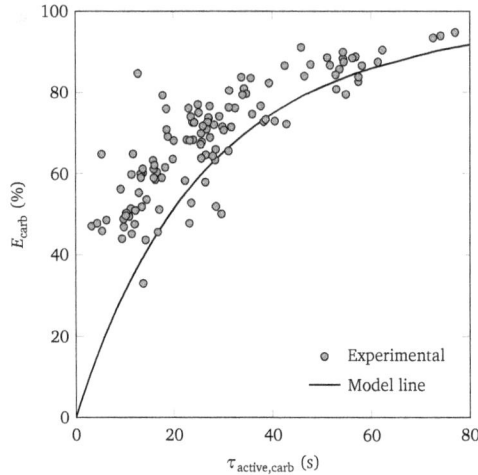

Figure 4.10: Carbonator absorption efficiency depending on the active space time.

As discussed by Rodriguez [104, 113], inherent uncertainties in the determination of the parameters applied in this model must not be neglected. The estimation of the solid circulation rate is challenging since it is a difficult parameter to be measured in a CFB system. Furthermore, the determination of average carbonate conversion is sensitive to experimental uncertainties since the extracted solid samples are only a very small share compared with the inventories and streams in the pilot plant. Nevertheless, the results depicted in Fig. 4.10 are in accordance with results of other pilots operated as CFB systems. The evaluation provides confidence for active space times $\tau_{active,carb} > 50$ s about the scalability in terms of carbonator absorption efficiencies $E_{carb} >> 80\%$.

4.4.4 Inventory

The influence of the carbonator inventory on the carbonator absorption efficiency corresponds to the impact of the active space time. The amount of Ca particles available for the carbonation reaction depends on the solid inventory. A high solid inventory in CFB systems increases the residence time of the particles to maximize the conversion but at the same time to cause a high pressure drop that leads to an increased power consumption of the fans. In contrast to the active space time, the evaluation inventory impact does not require a model approach and can be directly derived from the pressure measurement data.

A time course of specific carbonator inventory and absorption efficiency is shown in Fig. 4.11-a. This examination offers the advantage that all parameters such as sorbent properties, temperatures and flows were kept constant. Thus, it is possible to point out the effect of this single parameter. During the depicted period, the specific carbonator inventory $W_{s,carb}$ was decreased batch-wise with material extraction by a screw conveyor to keep it in a range of 600-800 kg/m^2. The impact on the carbonator absorption efficiency E_{carb} is obvious since every extraction leads to a decrease of efficiency that is reversed with increased inventory again. The dependency of the specific inventory on the absorption efficiency is degressive, i. e. at a certain value a continuous increase does not lead to a further increase of absorption. The residence time of the particles is long enough to reach the maximum conversion. The specific inventories of this period are plotted against the carbonator absorption efficiency in Fig. 4.11-b to show this effect. The scatter band of the measured inventories shows a stagnating efficiency at around 700 kg/m^2 in this case. The minimum inventory to realize appropriate absorption efficiencies depends on the CO_2 carrying capacity of the sorbent and the molar flow circulating between the reactors.

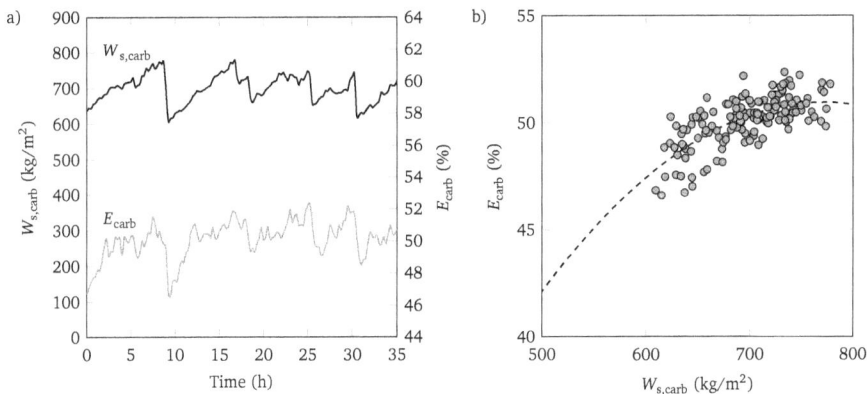

Figure 4.11: Influence of specific carbonator inventory on the absorption efficiency.

To illustrate the impact of the solid inventory on the carbonator efficiency for various operating points, the results for make-up ratios of 0.11 and 0.06 mole$_{Ca}$/mole$_{CO2}$ with sorbent looping ratios of 10-16 mole$_{Ca}$/mole$_{CO2}$ (MUR and LR averaged for considered operating points) are depicted in Fig. 4.12. Both Figs. 4.12-a and -b show that the CO_2 absorption efficiency rises with increasing specific inventories until a certain limit is asymptotically reached. The maximum efficiency limited by the equilibrium $E_{carb,eq}$ can only be reached if the flow of active Ca particles ($F_{Ca}X_{avg}$) is higher than the flow of CO_2 entering the carbonator, and if there is sufficient bed inventory to guarantee that most of the entering particles reach a conversion close to their maximum. If this is not the case and the carbonation capacity is not sufficient to absorb all CO_2 fed to the carbonator, the efficiency is limited by the sorbent looping ratio. A stagnating efficiency with increasing solid inventory thereby indicates the upper limit. The solids inventory is sufficiently high to ensure that most of the particles achieve their maximum conversion.

a) $\overline{MUR} = 0.11$ mole$_{Ca}$/mole$_{CO2}$ b) $\overline{MUR} = 0.06$ mole$_{Ca}$/mole$_{CO2}$

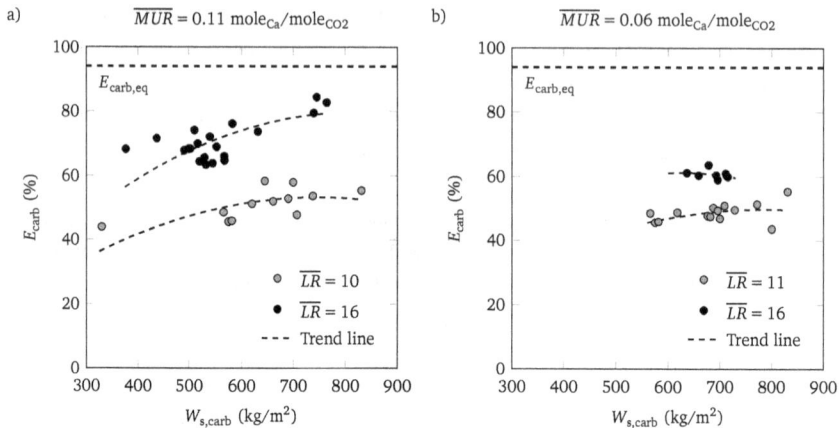

Figure 4.12: Carbonator absorption efficiency dependent on the specific solids inventory at different sorbent looping and make-up ratios.

As indicated in Fig. 4.12-a with a make-up ratio of 1.0 mole$_{Ca}$/mole$_{CO2}$ and a corresponding $\overline{X}_{avg}$ of 0.06 mole$_{CaCO3}$/mole$_{Ca}$, the upper limit is reached at an specific inventory of 700 kg/m^2 for a sorbent looping ratio of 10 mole$_{Ca}$/mole$_{CO2}$. With a higher looping ratio of 15 mole$_{Ca}$/mole$_{CO2}$, a specific solid inventory of 600 kg/m^2 is sufficient. However, the solids inventory required to guarantee a sufficient residence time for conversion depends on the available flow of active Ca particles.

To emphasize this effect, the results of operation with a lower make-up rate of 0.5 mole$_{Ca}$/mole$_{CO2}$ are shown in Fig. 4.12-b. The corresponding CO_2 carrying capacity $\overline{X}_{avg}$ of the sorbent is relatively low with 0.03 $_{CaCO3}$/mole$_{Ca}$ since less fresh make-up limestone enters the system. As can be seen, an increase of efficiency is independent of the specific inventory of 500-800 kg/m^2. An improved efficiency can only be achieved by increasing the sorbent looping ratio instead of the specific inventory.

4.4.5 Make-up Ratio

The make-up is the continuous feed of fresh limestone to diminish the effect of sorbent deactivation and to adjust the activity of the sorbent inventory. This parameter is of great importance for the steady long-term operation of the process since the sorbent activity is crucial for the CO_2 capture performance. The reliable assessment demands steady-state process operation, i. e. stable conditions of gas and sorbent phases. Due to the slow deactivation of the material, continuous operation requires periods from several ten hours up to days for steady-state sorbent conditions. To assure steady-state operation, the mean residence time of the sorbent was taken into account to estimate the required period to reach steady-state conditions in the solid phase. Three times the residence time of the sorbent in the system was utilized to ensure that at least 95 % of the inventory is exchanged by the make-up flow.

An exemplary long-term test run is shown in Fig. 4.13 to point out the influence of the make-up feed. The average temperatures in the carbonator and the calciner were adjusted to 650 °C and 870 °C, respectively, and were maintained during the entire test run. The carbonator was fluidized with coal-originated flue gas from the combustion chamber with a superficial gas velocity of 2.3-2.5 m/s. The specific reactor inventory was kept constant at 600 kg/m^2 during the whole testing period. Oxy-fuel combustion in the calciner was established using oxygen diluted with recirculated flue gas as fluidization media. The calciner was operated with a constant superficial gas velocity of 4.5-5.5 m/s and a lean zone temperature of around 900 °C to guarantee a full regeneration of the sorbent. The inventory was continuously replaced by a make-up feed of 0.45 mole/m^{2}s to guarantee a continuous mixing of inventory and fresh material, corresponding to a molar make-up ratio MUR of 0.11 mole $CaCO_3$ per mole CO_2 fed to the carbonator. The solid flow circulating between the reactors was kept constant at 3.8 kg/m^{2}s corresponding to a sorbent looping ratio LR of 17 mole of Ca circulated per mole CO_2 fed to the carbonator reactor for the absorption process.

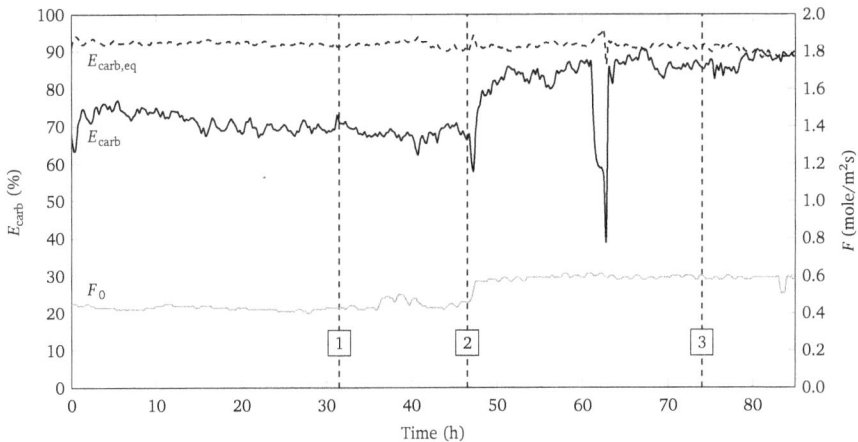

Figure 4.13: Long-term pilot operation and transition between two steady-states periods operated with different make-up ratios.

In the first phase, the absorption efficiency in the carbonator E_{carb} decreased from an initial value of 75 % to a residual value of 70 % while the established operating conditions were maintained for 46 hours to achieve a homogeneous mixing of inventory and make-up in order to guarantee steady-state sorbent behaviour. The decrease of E_{carb} corresponded to the reduction of the flow of CO_2 absorbed in the carbonator ($F_{CO2,carb,in}E_{carb}$) from an initial value of 2.9 to 2.5 mole/m^{2}s. After 31 h (1) steady-state operation was achieved. In this phase, steady-state conditions were recognizable by the constant CO_2 absorption in the gas phase. The solid analysis of three samples in the period from 31 to 46 h proved the steadiness in the sorbent.

After 46 h (2), the make-up feed to the process was increased from 0.45 to 0.6 mole/m^{2}s (MUR of 0.18 mole$_{Ca}$/mole$_{CO2}$) to adjust the sorbent performance, whilst the other operating parameters were kept constant. Fig. 4.13 shows that carbonator absorption efficiency increased significantly from 70 to 90 % in a period of 30 hours, representing conditions close to the thermodynamic equilibrium. The corresponding flow of absorbed CO_2 increased from 2.5 to 3.1 mole/m^{2}s. This operation period was shortly interrupted after 62 h when a combustion chamber shut-down forced carbonator fluidization by artificial flue gas for 30 minutes. During this period, less CO_2 and the absence of H_2O in the carbonator fluidization led to a decrease of T_{carb} to 615 °C, and as a consequence E_{carb} dropped. The transition phase ended after 30 h and after 72 h (3) a second steady-state operating period was reached representing stable conditions in gas as well as in the sorbent phase.

The results of sorbent analysis in both steady-state operational periods are shown in Fig. 4.14. As depicted in black, the sorbent is composed of 6.21 wt.% of $CaCO_3$, 1.45 wt.% of $CaSO_4$ and 6.06 wt.% of ash, balanced with CaO. The values correspond to a molar carbonate content of 0.038 mole$_{CaCO3}$/mole$_{Ca}$ and a molar sulphur content of 0.007 mole$_S$/mole$_{Ca}$. With increasing make-up rate, the transition of sorbent composition is reflected by the values in grey in Fig. 4.14. It is recognizable that the amount of $CaCO_3$ significantly increased to 9.28 wt.% and this reflects the higher loading of the circulating sorbent due to a higher activity of 0.057 mole$_{CaCO3}$/mole$_{Ca}$. At the same time, more inactive material was replaced. Thus, the fractions of $CaSO_4$ and ash decreased to 0.78 and 5.39 wt.%, respectively.

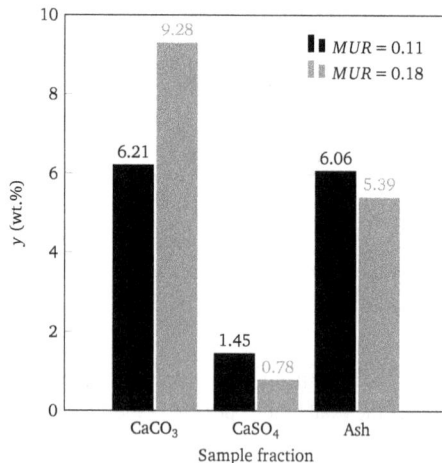

Figure 4.14: Sorbent composition of steady-state operating periods in Fig. 4.13.

As a consequence, the process was operated with a more active sorbent close to the thermodynamic equilibrium with a carbonator absorption efficiency of 90 %.

The previous paragraphs show the sensitivity of the process to changes of the make-up feed during long-term operation. The evaluation of different operating points for two specific inventories and sorbent looping ratios, respectively, are shown in Fig. 4.15. For the case of 640 kg/m² (see Fig. 4.15-a), the carbonator absorption efficiency increases with increasing make-up rate for both sorbent looping ratios. The lower the looping ratio, the more sensitive the carbonator efficiency is to changes in the make-up ratio. Fig. 4.15-b depicts the same correlation for a higher inventory of 760 kg/m² where you can see the same effect for the lower sorbent looping ratio. At the higher looping ratio, the make-up rate has no significant influence on the carbonator efficiency. As can be seen, high carbonator absorption efficiency is also possible with a very low make-up flow and a high inventory and sorbent looping ratio that compensate the low activity of the material.

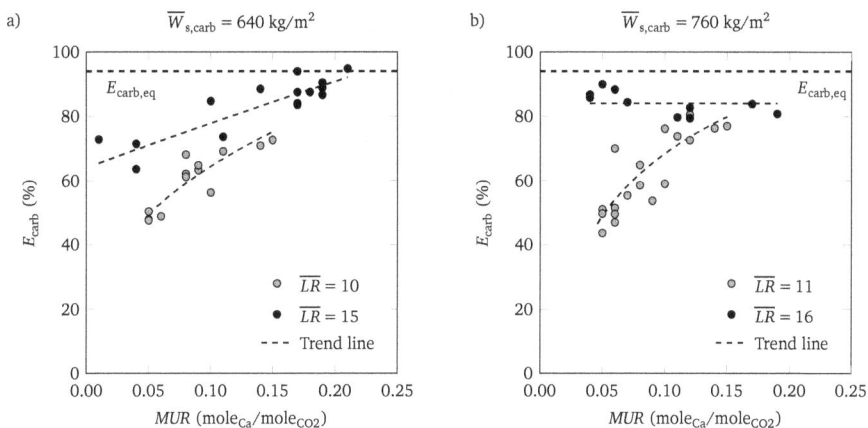

Figure 4.15: The effect of the make-up rate on the carbonator efficiency for two different specific inventories and sorbent looping ratios.

4.4.6 Sorbent Looping Ratio

As indicated in the sections before, the sorbent looping ratio is an important parameter of the CaL process. It is defined as the relation of the flow of Ca particles and CO_2 entering the carbonator. The ideal case is a sorbent looping ratio of 1, a mole of Ca absorbs a mole of CO_2 and full conversion takes place. Due to the cyclic deactivation of the sorbent, the maximum conversion is significantly lower. The optimum of the sorbent looping ratio is primarily determined by the CO_2 absorption capacity or CO_2 transport capacity, respectively, of the sorbent.

To achieve highly efficient carbonator absorption close to equilibrium conditions, the carbonation reaction must not be chemically limited. Therefore, the active space time $\tau_{active,carb}$ is an important parameter to describe the active inventory and the reaction rate of the particles in the carbonator. Thus, the influences of the specific inventory and the make-up ratio are considered to point out the influence of the sorbent looping ratio. In addition, this parameter strongly influences the heat input required in the calciner since the circulating particle stream has to be heated for calcination. Thus, it is important to achieve high carbonator efficiencies at low sorbent looping ratios.

The results from long-term pilot operation with low (15 s) and high (55 s) active space time is shown in Fig. 4.16. For low active space time, the inventories are around $600\,kg/m^2$ with an average make-up ratio of $0.09\,mole_{Ca}/mole_{CO2}$. Higher space time is achieved by inventories $720\,kg/m^2$ and make-up ratios around $0.13\,mole_{Ca}/mole_{CO2}$. The results show that the influence of the sorbent looping ratio decreases with increasing active space time. At lower space time, an increment of the sorbent looping ratio from 7 to 12 leads to a significant increase of the absorption efficiency, whereas the effect at 55 s is minor. In summary, high carbonator efficiencies can be achieved either by an increment of the active space time at a certain sorbent looping ratio or an increment of the sorbent looping ratio at a certain active space time. Furthermore, it can be observed that carbonator efficiencies close to the equilibrium require high hyperstoichiometric flows of active Ca particles.

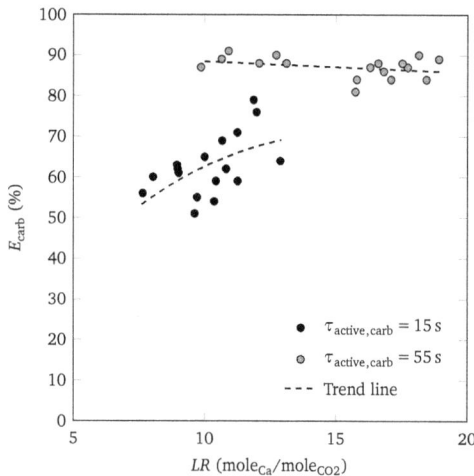

Figure 4.16: Carbonator efficiency over sorbent looping ratio for different the active space times.

4.5 Calciner Evaluation

The efficiency of the calciner is essential for the performance of the CaL technology. The regeneration efficiency is determined by a variety of parameters, such as the reactor temperature, the solids inventory or the sorbent properties. Many of these parameters correlate with each other and thus considerably complicate the evaluation of isolated effects. The following section discusses the different factors of influence on the calciner performance.

4.5.1 Active Space Time

The operation of a highly efficient carbonator requires calcined sorbent which is able to absorb CO_2. Thus, the calciner efficiency E_{calc} is responsible for the molar $CaCO_3$ fraction X_{calc} in/after the calciner. E_{calc} describes how many of the carbonated particles entering the calciner are regenerated when leaving the reactor (see Eq. 2.18). The conversion is influenced by the calciner conditions such as maximum temperature and CO_2 concentration. A molar carbonate content close to zero after calcination is important for the carbonator operation. Thus, a high regeneration of the sorbent minimizes the sorbent looping ratio and increases the possible CO_2 carrying capacity. Two main process parameters are responsible for the calcination efficiency [109, 139–141]. First, the residence times of particles in the calciner reactor t_{calc}, and second, the molar conversion of the sorbent X_{carb}. The residence time considers the reactor design. The molar carbonate conversion depends on the operating conditions of the carbonator, and represents the sorbent properties. Both parameters lead to the active space time of the calciner, as shown per Eq. 4.9. It describes the inverse calcination load, i. e. the amount of sorbent to be calcined in a certain time.

$$\tau_{active,calc} = \frac{n_{Ca,calc}}{F_{Ca}X_{carb}} = \frac{t_{calc}}{X_{carb}} \tag{4.9}$$

The calciner efficiency and the molar carbonate content at the calciner outlet as a function of the active calciner space time is shown in Fig. 4.17. The considered operating points are divided by the different calciner reactor setups, once-through (see Fig. 3.2-a) and with internal solid recirculation in the reactor loop (see Fig. 3.2-b). In both considered setups, the calciner conditions were kept in a comparable range. The calciner average temperatures were around 850 °C with maximum temperatures in the upper part of at least 920 °C. The CO_2 outlet concentrations were 60-70 vol.% and the entering molar carbonate content 0.04-0.06 $mole_{CaCO3}/mole_{CO2}$. The once-through calciner was operated with very short residence times between 12-30 s, whereas the internal solid recirculation slighty increased the residence time up to 50 s.

The general trend in Figs. 4.17-a and -c shows increasing calciner efficiency with decreasing molar carbonate content at the calciner exit for the operation without internal solid recirculation. The once-through calciner achieves very high efficiencies > 95 % partially at a very low active spaces time of 3 min. With an active space time > 10 min, the residual calciner efficiency is very close to 100 % (see Fig. 4.17-a). The corresponding molar carbonate content of the solid transferred to the calciner decreases from 0.017 to 0.001 $mole_{CaCO3}/mole_{CO2}$ (see Fig. 4.17-c). The results show that both E_{calc} and X_{calc} correlate with the active space time. Additionally, it is important to know that a once-through calcination under oxy-fuel conditions works with very high efficiencies.

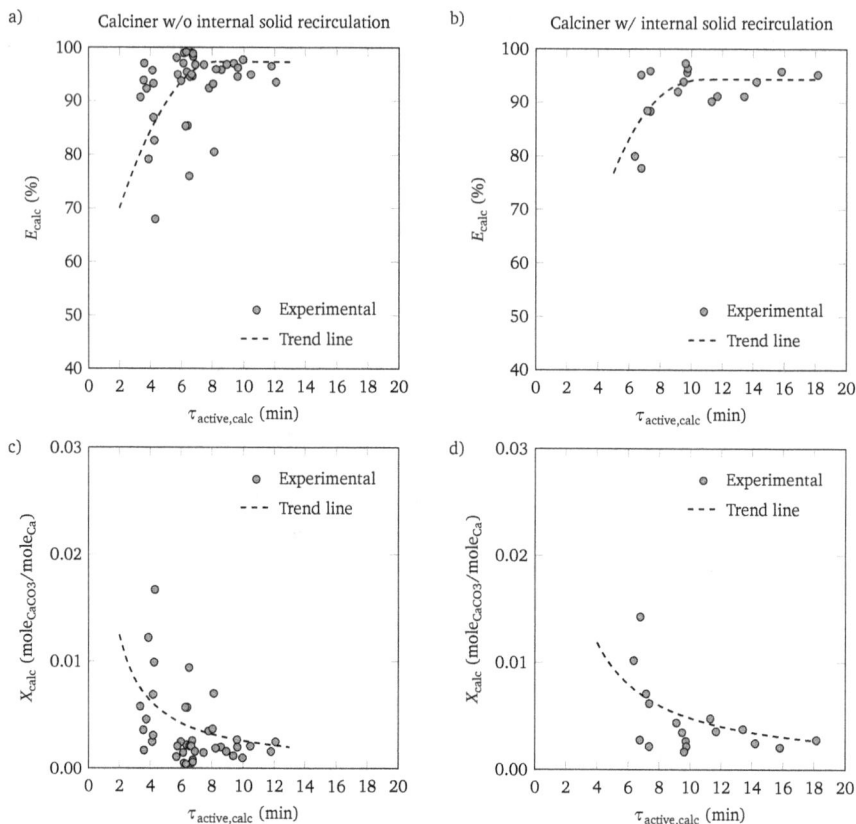

Figure 4.17: Calciner efficiency and residual carbonate content dependent on the active space time for reactor setups with (left) and without (right) internal solid recirculation.

Operating the calciner with internal solid recirculation shows the same trends. By achieving an active space time of 7 min, calciner efficiencies $> 90\,\%$ are possible. Efficiency close to $100\,\%$ can be achieved by 13 min or more (see Fig. 4.17-b). The residual molar carbonate content decreases from 0.014 to $0.002\,\text{mole}_{CaCO3}/\text{mole}_{CO2}$ with increasing active space time (see Fig. 4.17-d). This concept also offers the possibility to efficiently run the calcination in the CaL process.

The conclusion can be drawn that both concepts are feasible for operation with high calcination efficiencies. A sorbent with a low molar carbonate content is provided for the carbonator. The comparison of both concepts shows a difference in the calcination efficiency at a given active space time. There is the tendency that the once-through calciner achieves higher calcination efficiencies at lower space times at almost similiar operating conditions. By comparing the residual molar carbonate content, the once-through calciner delivers lower values down to $0.001\,mole_{CaCO3}/mole_{CO2}$. The particles in the other concept are partially recirculated and have a slightly higher residual of $0.002\,mole_{CaCO3}/mole_{CO2}$. As an explanation of this effect, it is taken into consideration that the calciner temperatures in the bottom zone are below 700 °C and the calciner particles re-enter the bottom zone. There, CO_2 concentration of 30-40 vol.% are pre-dominant due to the flue gas recirculation. The particles partially recarbonate since the bottom zone temperature is lower than the equilibrium temperature of 820-840 °C. Thus, the material stream can not be fully converted and some residual carbonate content is left in the sorbent transferred to the calciner. The once-through calcination avoids this effect because all calciner particles are directly routed to the carbonator.

4.5.2 Temperature and CO_2 Concentration

On the efficiency of the calciner, the temperature and the CO_2 concentration have a significant impact in addition to the influence of the active space time. Both parameters are closely linked by the thermo-dynamic equilibrium. The calciner operation requires high temperatures since high CO_2 concentrations are predominant and undesired dilution of the CO_2 product needs to be avoided. Even with some recirculation gas and depending on the moisture of the fuel, CO_2 concentrations lower than 60 vol.% are not expected. Thus, the corresponding equilibrium temperature is already 862 °C. To keep a sufficiently fast calcination of the sorbent, higher temperatures than the equilibrium are required. If the temperature is too low the calcination reaction slows down too much and the equilibrium CO_2 concentration is lowered. Thus, the prevailing CO_2 concentration could exceed the equilibrium concentration and undesired recarbonation takes place. If the calciner efficiency is reduced, the carbonator cannot obtain sufficient quantities of calcined sorbent leading to a reduction of the CO_2 capture efficiency of the process.

The calciner efficiencies for different average calciner CO_2 outlet concentrations as a function of the maximum reactor temperature is shown in Fig 4.18. The maximum calciner temperature is chosen since the temperature profile of the reactor is inhomogeneous due to the cold sorbent and coal entering in the bottom zone. The fact that the calcination reaction of partially carbonated sorbent runs very fast and is mainly temperature dependent with sufficient residence time in the reactor [142], the maximum calciner temperature is used for evaluation. The required equilibrium temperature for CO_2 outlet concentration of 43 vol.% is 845 °C. As depicted in Fig. 4.18, the calcination efficiency at temperatures close to this value is low. With increasing difference between calciner reactor and equilibrium temperature, calcination efficiency increases. As the calciner temperature exceeds 890 °C, calcination efficiencies of 90 % are achieved. As expected, the minimum calciner temperature required rises with increasing CO_2 outlet concentration of 66 vol.% at a corresponding equilibrium temperature of 870 °C. To achieve calciner efficiencies of 90 %, temperatures of 925 °C or higher are beneficial. Exceeding 925 °C, the calciner efficiency remains above 90 %. In this way, working at a temperature level of at least 925 °C in the calciner, negative effects on the calcination efficiency will be diminished if there is any change in the CO_2 concentration in the reactor, e.g. a change of the fuel type or the flue gas recirculation rate.

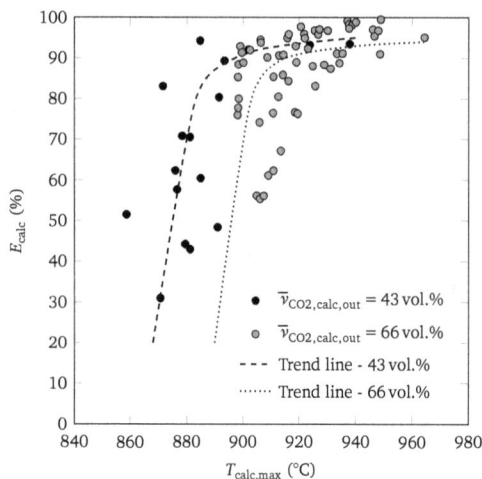

Figure 4.18: Calciner efficiency dependent on the maximum calciner temperature for different CO_2 outlet concentrations.

4.5.3 Fuel Type

Post-combustion CO_2 capture by CaL has rapidly developed for coal combustion applications. As a consequence of the nature of the process, the continuous feed of fuel raises the challenge of inert solid accumulation. This effect is mainly evoked by the components of the coal fed to the calciner. The sulphur reacts with the CaO of the sorbent inventory to form $CaSO_4$ and ash accumulates. Accumulations of inerts can be prevented by establishing a continuous purge. The accumulation of impurities depends on the ratio of inerts fed to the system and the feed of make-up added to the process. The inert feed strongly depends on the composition of the fuel fed to the calciner for sorbent regeneration. Both parameters determine the composition and the particle size distribution of the process inventory affecting the efficiency, heat demand and the operability of the process.

The effect of the fuel type was examined by the variation of the fuel in the test campaigns. Results of sorbent analysis of samples from pilot tests with pulverized hard coal (HC1) and lignite (LEP) are depicted in Fig. 4.19-a. The graph depicts a significant difference in the content of $CaSO_4$. In the hard coal case, 11 wt.% of $CaSO_4$ is derived from the analysis, whereas only 1 wt.% is the result in the lignite case. The low sulphur content in the lignite of 0.19 g/MJ compared to the hard coal with 0.26 g/MJ significantly affects the accumulation of calcium sulphate reducing inactive material in the process. The comparison of ash analysis in Fig. 4.19-a shows no significant difference in the accumulation of ash in the sorbent sample, although the ash content in the lignite of 2.49 g/MJ is lower than 5.87 g/MJ of the hard coal. The mass ratios of make-up to ash fed to the process in these operating points were 2.5 (hard coal) and 8 (lignite), respectively. It is assumed that only a certain particle fraction of ash is separated by the cyclone independent of the fuel type. For both cases, the amount of ash particles in the separated fraction in the cyclone is the same, whereas the rest of ash particles left the reactor via cyclone and was retained in the bag filter. The analysis of filter samples for the hard coal case show ash fractions of 35 wt.% for the carbonator and 41 wt.% for the calciner. The $CaCO_3$ fractions of both filter samples were around 28 wt.%. In the case of lignite, the ash share in the filters were lower with 9 wt.% for the carbonator and 19 wt.% for the calciner, respectively. The $CaCO_3$ fractions for this case are in the range of 11 to 15 wt.%.

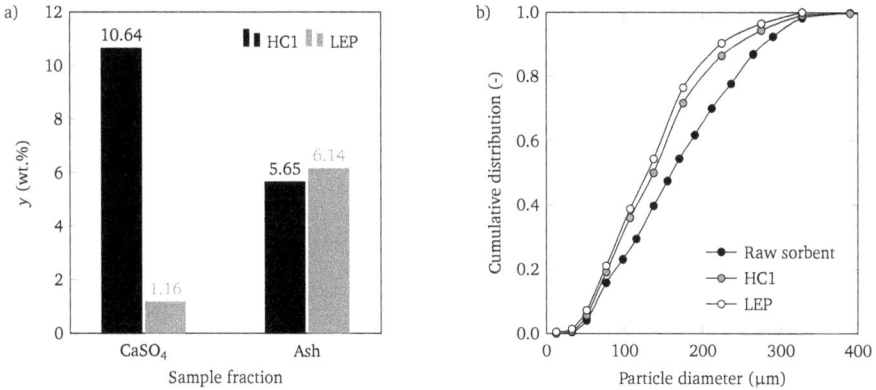

Figure 4.19: Sorbent composition (a) and particle size distribution (b) from calciner operation with pulverized fuel (hard coal and lignite).

Another important subject for CaL design and scale-up is the development of the PSD of the sorbent. Mechanical behaviour and composition influence the evolution of particle size over the time. The particle size distributions of raw material and calciner loop seal samples from steady-state operation are shown in Fig. 4.19-b. The initial median particle size of the raw material $d_{50} = 180\,\mu m$ is decreased to 138 μm for pulverized hard coal and to 129 μm for pulverized lignite. The decrease corresponds with the reduction of the initial particle size of 21 % and 26 %, respectively. The phenomenon can be explained by two effects, the accumulation of fine ash and the breaking of particles. The main effect is assigned to the second one. The sorbent particle edges and asperities are knocked off so that the particles become smaller, rounder and smoother with progressing time until a steady-state condition is reached. This size reduction is considered independent of the fuel type since both samples have a similar share of ash. For all samples low shares of fines $< 50\,\mu m$ in the range from 5 to 7 wt.% were observed. Fines from attrition and the make-up stream were segregated and exited the system with the fly ash. Thus, the hydrodynamic stability of the CaL system was not negatively influenced.

To diminish the effect of accumulation of inerts, the make-up feed is crucial for a given fuel. Therefore, the mass fraction of inert material (y_{CaSO4}, y_{ash}) in the stream of calcined particles transferred from calciner to carbonator is shown in Fig. 4.20. The results of operation with two types of pulverized hard coal and lignite are depicted against the make-up rate MUR. The trend shows decreasing accumulation with increasing make-up feed. A very low make-up feed leads to a strong increase of accumulated impurities. Increased make-up rates show a trend up to a residual value and depend on the fuel composition. A further increase of the make-up rate does not lead to a significant lowering of the accumulations. As HC1 has a high sulphur and ash content with 0.26 g/MJ and 5.87 g/MJ, respectively, a very high accumulation of 21 wt.% with low make-up ratios of 0.05 mole/mole is obvious. It can be lowered to 15 wt.% by increasing the make-up ratio to at least 0.1 mole/mole. The same trend at different levels of accumulation can be seen for the other types of coal as well. The same sulphur and a lower ash content of HC2 with 0.29 g/MJ and 2.92 g/MJ lead to significant lower accumulation of 15 wt.% with low make-up ratios. The increase of make-up leads to residual accumulations of 10 wt.%. A very low share of accumulations can achieved with a low sulphur and ash fuel, such as the pulverized lignite LEP. The sulphur content of 0.19 wt.% and the ash content of 2.49 wt.% are very low. This high quality is reflected by the results. For almost the complete range of make-up ratios, the inert share was lower than 9 wt.%.

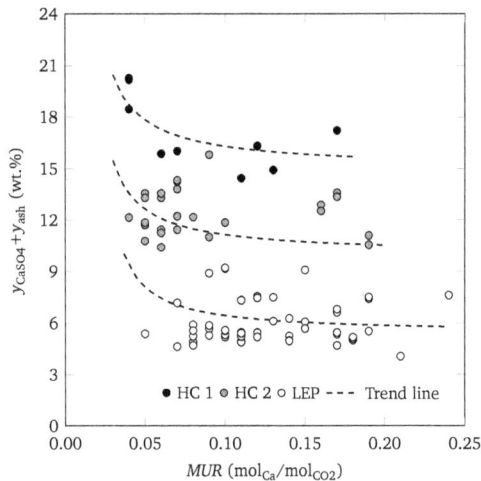

Figure 4.20: Mass fraction of accumulated inert material in the solid stream transferred from calciner to carbonator dependent on the make-up ratio for various types of fuel.

To face the challenges of scale-up, the influence of the fuel type has to be considered in the design of the plant. A certain make-up ratio is not only required to achieve the desired activity of the material, it is also important for the composition of the inventory and the circulating sorbent. A circulating sorbent stream with a high share of impurities causes unnecessary heat losses. An increase of the make-up ratio is feasible to limit the inert share to a residual value.

Moreover, the specific surface and the pore volume as characteristic sorbent properties were evaluated with respect to the utilized fuel type. The specific surface was determined with Brunauer-Emmet-Teller-Analysis (BET) and the pore volume with Barrett-Joyner-Halenda-Analysis (BJH). Both parameters show the influence of the heating on the sorbent. The operation with hard coal shows specific surfaces of 1.1-2.3 m^2/g whereas operation with lignite leads to significantly lower values of 0.4-0.8 m^2/g. These values are in accordance with findings in literature [77, 82, 143]. To point out these differences, Table 4.4 shows the surface, pore volume and in addition the molar carbonate content from samples taken at the carbonator outlet. The selected and shown samples are taken from a hard coal and a lignite test campaign with the same sorbent (MF).

The results indicate a significant influence of the fuel type on the specific surface. However, molar conversion rates in the same order of magnitude are achieved with both fuels (see Table 4.4: hard coal-HC, lignite-LEP) although there is a difference in specific surface and pore volume. The influence of sulphur or the sintering by the reactor temperature could be possible explanations for the BET and BJH values. Since all tests were carried out at similar temperatures but with different fuel types, the combustion behaviour of each fuel could explain the differences. Lignite has a higher share of volatiles leading to higher local temperatures. As a consequence, this type of fuel promotes sintering. In addition, the fact that the significantly less sulphated sample from the lignite campaign has a lower specific surface area than the heavily sulphated sample from the hard coal campaign does not support the assumption of the sulphur influence. The question why the specific sorbent surface of samples from lignite tests is much smaller than the ones with hard coal, cannot be finally clarified within this work since the number of samples evaluated is only limited. Additional measurements and evaluation methods such as the extensive microscopic analysis are required in order to achieve a reliable assessment.

Table 4.4: Specific surface, pore volume and molar carbonate conversion for chosen sorbent samples from operation with hard coal (HC) and lignite (LEP).

Description	BET surface	BJH pore volume	X_{carb}
-	m^2/g	cm^3/g	mol$_{CaCO3}$/mol$_{Ca}$
HC-1.116	1.32	3.5 10^{-3}	0.050
HC-1.212	1.33	4.1 10^{-3}	0.047
HC-1.271	1.33	3.8 10^{-3}	0.051
HC-1.277	1.32	3.6 10^{-3}	0.052
LEP-2.100	0.43	1.0 10^{-3}	0.041
LEP-2.118	0.45	1.0 10^{-3}	0.038
LEP-2.127	0.55	1.4 10^{-3}	0.039
LEP-2.173	0.73	1.7 10^{-3}	0.057
LEP-2.208	0.79	1.9 10^{-3}	0.059

As part of the investigation of the fuel flexibility of the CaL process, fuels with different particle size distributions were fed to the calciner. For this, the same type of fuel with different particle size distributions was utilized for sorbent regeneration. The Colombian hard coal HC1 was used either in coarse ($d_{50} = 1{,}500\,\mu m$, sieved with a maximum size of $10\,mm$) or in pulverized ($d_{50} = 45\,\mu m$) particle size in the calciner. Operating points from test campaigns utilizing the same sorbent with these different fuel preparations are compared to point out the differences in sorbent composition and PSD. In these steady-state operating points compared here, the make-up ratio was equal. Solid samples representing the sorbent composition of the calciner loop seal were analysed and compared.

The results from solid analysis can be read from Fig. 4.21-a. The analysis show a similar share of 10-11 wt.% $CaSO_4$ for both samples, i. e. the particle size of the fuel does not influence the accumulation of sulphur in the sorbent. As expected, the sulphur is released to the gas phase during devotilization and combustion. In a comparison of the ash content of the samples, a significant difference is obvious. The sample from operation with coarse coal shows approx. 11 wt.% of ash compared to only 6 wt.% in samples from pulverized coal. This trend was observed during many operating periods independent of the make-up ratio. During operation with coarse hard coal, an ash fraction of 10 to 18 wt.% accumulated in the circulating material. This leads to a conclusion that the particle size of the coal significantly affects the share of ash. The separation of ash particles in the cyclone is affected. The bigger the ash particles, the more ash is separated from the calciner off-gas and accumulates in the circulating material. The consequence for the design of the process when firing coarse coal in the calciner is an increased purge stream to keep the ash accumulation low.

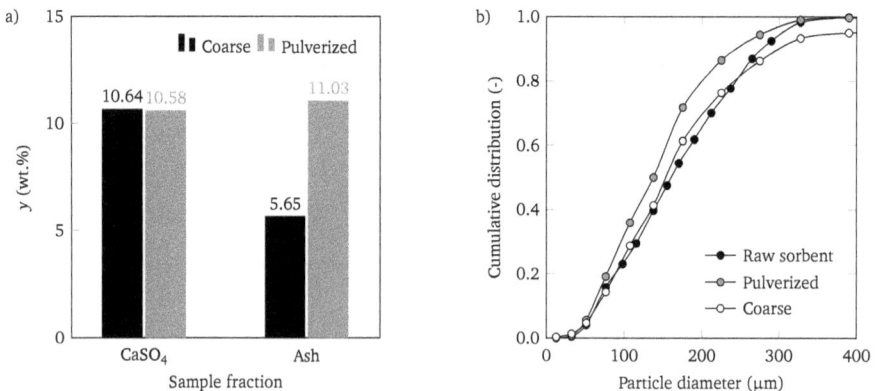

Figure 4.21: Sorbent composition (a) and particle size distribution (b) from calciner operation with coarse and pulverized hard coal as fuel.

To evaluate influence of mechanical behaviour and composition on the evolution of particle size over the time, the PSDs of raw material and calciner loop seal samples from steady-state operation are shown in Fig. 4.21-b. The initial median particle size of the raw material, $d_{50} = 180\,\mu m$, was decreased to $138\,\mu m$ for pulverized hard coal and to $154\,\mu m$ for coarse hard coal. The significantly higher fraction of ash in the case of coarse (11 wt.%) compared to pulverized (6 wt.%) coal increases the median particle size diameter. The separation of bigger sized ash particles is shown by an accumulation of approx. 10 % of ash particles > 400 μm. For all samples low shares of fines < 50 μm in the range from 5 to 7 wt.% were observed. Fines from attrition and the make-up stream exited the system with the fly ash. Thus, the hydrodynamic stability of the CaL system was not negatively influenced.

So far the challenge of the coal combustion in the calciner has only been dealt with by Shimizu et al. [144] in laboratory scale. The carry-over of unburned char to the carbonator needs to be addressed in order to develop the technology for industrial application. If unburnt char enters the carbonator, oxidation of char occurs in the reaction with the oxygen in the flues gas that has to be de-carbonized. Either CO could be formed in the carbonator and released to the atmosphere without being oxidized, or CO is oxidized to CO_2 or char is directly oxidized to CO_2. Thus, the formed CO_2 will react with the CaO sorbent in the carbonator causing an additional supply by the calciner. If the absorption is limited, the CO_2 concentration of the decarbonized flue gas will increase.

Various observations were made during operation with coarse hard coal in the calciner that indicate the effects of different fuel particle sizes fed to the calciner. Fig. 4.22-a shows the effect of a slight increase of the fuel feed during steady operation. As a consequence, the carbonator O_2 outlet concentration drops. This is an indicator for oxidizing reaction, such as the formation of CO by oxidizing char. The change from pulverized (d_{50} ̄0 μm) to coarse (d_{50} ̄00 μm) fuel is shown in Fig. 4.22-b. The CO concentration increases while the O_2 concentration at the carbonator outlet decreases when the fuel feed in the calciner is changed. Thus, the combustion efficiency in the calciner is influenced by the char burnout. The

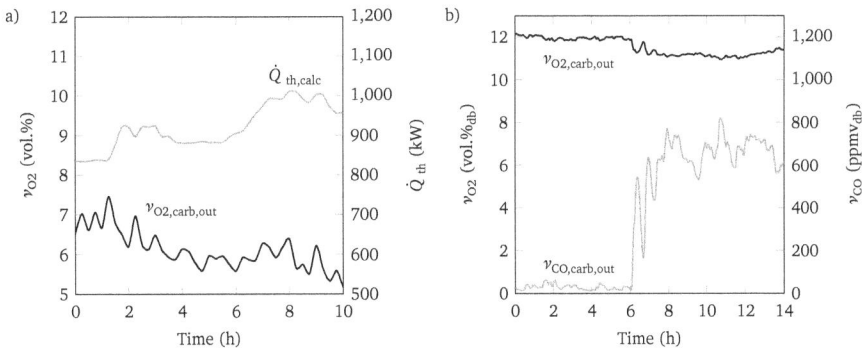

Figure 4.22: The effects of firing coarse coal in the calciner: variation of fuel input (a) and transition from pulverized to coarse hard coal (b).

burnout is calculated by mass balance and is lower for coarse coal (65-70%) compared to the pulverized fuel (85-90%) under the tested operating conditions since the residence time is too short to completely burn the char. The conclusion is that a part of devolatilized and partially burned light char particles is entrained, separated by the cyclone, and transferred to the carbonator.

To further evaluate the observations made during operation, the balances for O_2, CO_2 and CO for both types of calciner fuel were examined to verify this effect. Fig. 4.23-a shows the oxygen balance of the carbonator. The values for operation with pulverized calciner fuel shows a very good agreement of in- and outgoing oxygen mass flow. Thus, the conclusion is drawn that no char is transferred from calciner. For operating points with coarse fuel, the balance shows a clear mismatch between outgoing and entering oxygen mass flow. Oxygen is consumed in oxidation reaction. Further, the CO balance is considered for evaluation, as depicted in Fig. 4.23-b. For pulverized coal, the balance shows a very low flow of incoming CO or none at all. The outgoing flows are either equal to the entering flow of CO or were reduced by passing the carbonator, as postulated by Shimizu et al. [144]. The values of coarse coal operation stand in contrast to this results. The balance clearly indicates the reaction of char with oxygen to CO in the carbonator. The O_2 consumption and the CO formation are used to calculate the additional CO_2 released in the carbonator.

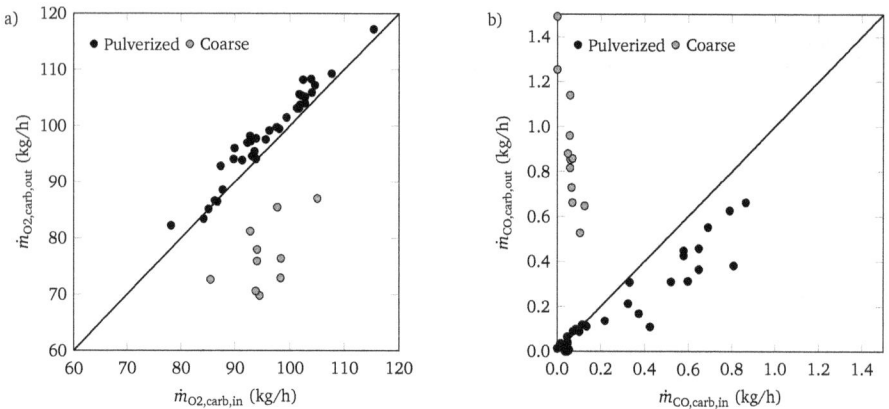

Figure 4.23: Oxygen (a) und carbon monoxide (b) balance of the carbonator for operation with pulverized and coarse calciner fuel.

The balances of O_2 and CO are used to evaluate the effects on the carbonator. It is supported by a more deeper analysis including the carbon content of solid samples exiting the calciner. The sorbent analysis shows a carbon content of the circulating stream of approx. 0.24 wt.%. Depending on the considered operating point, the oxidization reactions in the carbonator releasd 40-120 kW of additional heat and 15-47 kg/h of CO_2. Thus, 8-23% more CO_2 needed to be absorbed. Since the cooling capacity of the carbonator was limited, the solid circulation had to be decreased to keep the temperatures within the desired range. As a consequence of the limited supply with calciner material and the additional CO_2 resulting by char oxidation, the carbonator absorption efficiency related to the flue gas entering the reactor was limited.

The described effects of the fuel particle size in the calciner have to be considered in the engineering of a scaled-up calciner reactor. To avoid coal transfer from calciner to carbonator, either the residence times of coarser coal particles have to be increased by reducing the superficial gas velocity, or the fuel with a smaller PSD that combusts faster has to be utilized.

4.5.5 Heat Demand

The heat and corresponding oxygen demand of the CaL process, i. e. the oxy-fuel firing in the calciner, is determined by various parameters. The fuel input is of importance for the technological and economic feasibility of the process. The heat demand is mainly affected by the CO_2 absorption efficiency in the carbonator, the sorbent looping ratio, the temperature difference between the reactors, the fluidization flow of oxygen and recirculated flue gas as well as the required feed of make-up.

Table 4.5 shows the consumption of the different heat flows of a steady-state operating point related to the fuel input into the calciner. Based on the actual combustion heat of the coal, the partial heat flows and their percentage are shown. For this operating point, almost 75 % of the heat is consumed by the heating of the circulating sorbent and the fluidization gas to calcination temperature. Only 18 % of the heat is required for calcination of sorbent and make-up. Hereafter, the heat requirements during the pilot tests are assessed and the influence of carbonator efficiency, sorbent heating and make-up feed is evaluated. The flue gas losses with respect to the oxygen concentration and the related flue recirculation rate is not assessed since the operating window was too small to point out the influence. Two test campaigns with the same fuel pulverized lignite (LEP 1 and LEP 2) are selected to show the trends observed during pilot operation.

Table 4.5: Calciner energy balance for a steady-state operating point with 662 kW$_{th}$ fuel input.

Heat flow	Absolute	Relative
Heating of circulating sorbent	298 kW$_{th}$	45 %
Flue gas loss	189 kW$_{th}$	29 %
Calcination of sorbent	108 kW$_{th}$	15 %
Calcination of make-up	17 kW$_{th}$	3 %
Heat losses	50 kW$_{th}$	8 %

The heat ratio is used to compare the results independent of the carbonator input. As shown in Eq. 2.21, it is defined as the ratio between the required heat input to the calciner to run the process and the thermal equivalent of flues gas entering the carbonator. The ratio should be as low as possible to decrease the fuel and oxygen input to the calciner, and consequently the costs. Complex heat transfer between the different streams, especially for the circulating sorbent between the reactors, significantly lower the heat ratio. This complexity is not represented in the basic setup of the pilot plant.

An overview of the achieved heat ratios as a function of the flue gas heat equivalent of the carbonator fluidization during pilot operation is shown in Fig. 4.24. The flue gas entering the carbonator is within the range of 370-790 kW_{th} while the calciner heat input is 630-1,180 kW_{th}. The data shows a considerable spread with heat ratios of 1.2-2.8. In general, the required fuel input clearly exceeds the carbonator heat equivalent. A factor to be considered is the incomplete burnout of the coal. Heat and mass balancing show that a part of the fuel input (approx. 10-15 %) is not fully oxidized and the heat could not be utilized for sorbent regeneration. The main influencing factors on the heat ratio of the CaL process do not usually appear isolated, but in different combinations complicating the analysis.

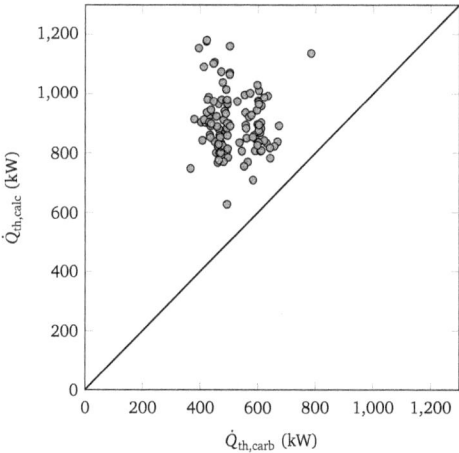

Figure 4.24: Comparison of thermal carbonator flue gas equivalent and the fuel input of the calciner.

The carbonator absorption efficiency directly influences the heat ratio. The more CO_2 is captured in the carbonator and is bound in the circulating stream, the more fuel is required to regenerate the sorbent. The heat ratios for the test runs with LEP fuel are depicted in Fig. 4.27 to show this effect. Achieving high carbonator absorption efficiencies of 80 % requires heat ratios of 1.8-2.5. Whether the increased heat requirement is entirely due to the increased CO_2 absorption in the carbonator, or whether the make-up and the sorbent looping ratios are also involved, cannot be answered in isolation.

The sorbent looping and the temperature difference between the carbonator and calciner represent a significant influence on the heat ratio. With increasing solid flow the required heat input rises since the material has to be heated from the colder carbonator to the warmer calciner level. Both the parameters of sorbent looping ratio and temperature difference between the reactors are considered in Fig. 4.26. It shows the heat ratio as a function of the temperature corrected sorbent looping ratio for two test

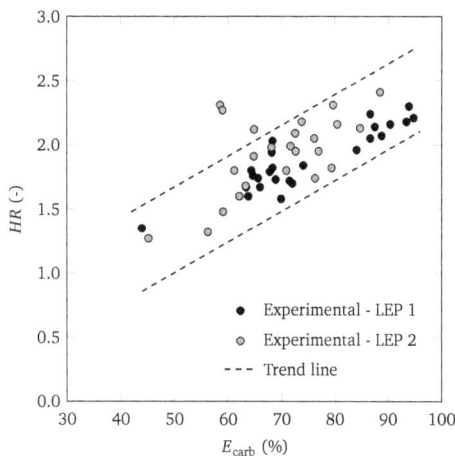

Figure 4.25: Heat ratio as a function of the carbonator absorption efficiency.

campaigns with pulverized lignite. The correction implies the multiplication of the sorbent looping ratio by a dimensionless parameter. It expresses the mean temperature difference between the carbonator and the calciner divided by the design temperature difference of 250 °C. The corrected sorbent looping ratios shown here correspond to sorbent looping ratios of 7-20 mole/mole and mean temperature differences of 120 to 240 °C. The effect of the sorbent looping ratio on the heat ratio is clearly visible in both test

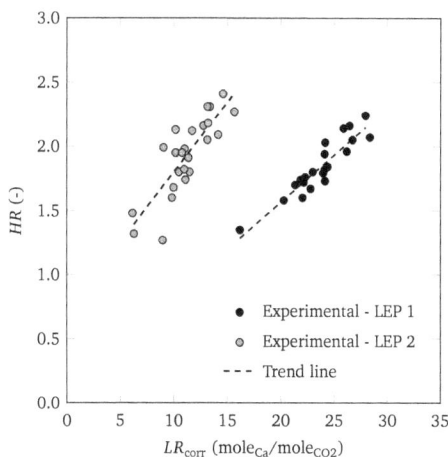

Figure 4.26: Heat ratio as a function of the temperature corrected sorbent looping ratio.

runs (see Fig. 4.26), even though there is a difference between both data sets. The results is influenced by make-up feed and carbonator absorption efficiency, but the trend is visible.

$$LR_{corr} = LR\frac{T_{calc} - T_{carb}}{250\,°C} \tag{4.10}$$

The make-up fed to the process needs to pass a first calcination. The additional heat input for this process is required in the calciner, so that the overall thermal input increases. The influence of the make-up ratio on the heating ratio is shown in Fig. 4.27. The correlation of both parameters is visible, even though there is some fluctuation. As already mentioned for the other parameters, the interdependence between the parameters explain the fluctuations.

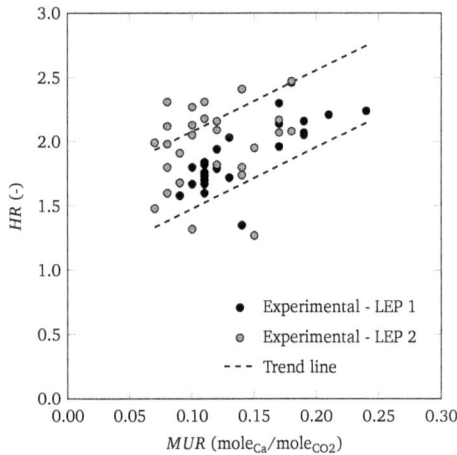

Figure 4.27: Heat ratio as a function of the make-up ratio.

4.6 Conclusions of Pilot Tests and Optimization Potentials

The previous sections have shown the experimental results obtained from the CaL process investigations in the 1 MW$_{th}$ pilot plant. The results included the assessment of the operability and the key parameters for the process performance, e. g. reactor temperatures and inventories, make-up and sorbent looping ratio, the fuel type and particle and the interdependencies with respect to process scale-up. A summary of the main achievements is given in the following section.

The long-term operability of the CaL process was proven with achieving steady-state conditions in gas and sorbent phases. In order to achieve steady state operation with special focus to the sorbent, the process was operated for periods of up to 60 hours without any changes of the control parameters to realize a continuous mixing of existing inventory and make-up stream. This procedure resulted in the assessment of the deactivation and the residual molar Ca conversion of the sorbent under realistic operation conditions for commercial application (coal originated flue gas, high CO_2 concentrations and temperatures in the calciner). The required proof of long-term operability under stable hydrodynamic and temperature profiles was realized with different reactor coupling concepts. On the one hand, a reactor interconnection with two loop seals, and on the other hand, a loop seal equipped with a cone valve were applied to transfer the solids from calciner to carbonator.

The operational experience and the results in $1\,MW_{th}$ scale provide the basis to define the operating window for a scaled-up demonstration or commercial plant. Carbonator absorption rates of $> 90\,\%$ corresponding to total CO_2 capture efficiencies of $> 95\,\%$ with different natural limestones and various calciner fuels under a wide range of operating parameters were achieved. These high CO_2 absorption rates can be realized by applying the highly relevant operating parameter carbonator temperature ($650\,^\circ$C to $675\,^\circ$C). Active space times $> 50\,$s are feasible to reach the targeted absorption rates but should be kept as low as possible to minimize parasitic losses. To accomplish high efficiency and to adjust the active space with a molar Ca conversion of 0.04-$0.06\,mole_{CaCO3}/mole_{Ca}$, conservative parameters of a specific inventory in the carbonator $> 800\,kg/m^2$, a make-up ratio $> 0.15\,mole_{Ca}/mole_{CO2}$ and a sorbent looping ratio of 10-$15\,mole_{Ca}/mole_{CO2}$ are foreseen as the basis to scale-up the process.

Furthermore, the calciner performance was assessed under oxy-calcination conditions with CO_2 concentrations up to $80\,.\%$. The results indicate that a calciner temperature of at least $925\,^\circ$C and an inventory of 60-$350\,kg/m^2$ are sufficient for high calcination efficiencies for regenerating sorbent which carries 0.04-$0.06\,mole_{CaCO3}/mole_{Ca}$. Exceeding an active space time for the calciner of $10\,min$, leads to calcination efficiencies $> 90\,\%$. Additionally, the sorbent composition is significantly influenced by the fuel feed into the calciner. The pilot tests with different types of fuel and varying particle size distributions show the accumulation of impurity fractions between 4 and $20\,wt.\%$. Thereby, the sulphur content of the coal has a significant impact on the calcium sulphate accumulation for pulverized fuels, whereas the fuel particle size significantly influences the accumulation of ash. The bigger the particle size of the ash, the more ash is separated from the cyclone and kept in the system. Less fine ash leaves the system and is retained in the bag filter.

The impact factors on the heat demand required to operate the process was derived from pilot operation. Heat ratios between 1.2 and 2.8 resulted from the pilot tests and the main influence factors were assessed. Most of the heat is consumed by heating up the circulating solid to calcination temperatures whereas the carbonator absorption efficiency and the make-up ratio have a minor impact on the heat demand of the process.

To quantify the heat demand per kilogramme of CO_2 captured in the process, the specific heat consumption Q_s is applied. Thus, the process is comparable to others. For the CaL pilot test, it is in a range of 6.2-$8.6\,MJ/kg_{CO2,capt}$. Compared to MEA with approx. $4\,MJ/kg_{CO2,capt}$ [145], the CaL process requires more heat input to capture the same amount CO_2. However, due to the high temperature level of the process, the heat can be efficiently utilized to compensate the comparably higher heat demand. Additionally, it is noteworthy to mention that the heat losses in the $1\,MW_{th}$ are significantly higher compared to larger size reactor systems and thus will decrease with increasing scale. The heat consumption of the directly fired CaL process is closely linked to the oxygen demand for oxy-fuel calcination. During pilot operation, $0.2\text{-}0.5\,kg_{O2}$ were required to capture a kilogramme of CO_2. In addition, make-up is consumed to control the sorbent activity that is required for efficient CO_2 absorption. Between 35 and 215 kg of fresh limestone were fed during operation in order to capture a ton of CO_2. High make-up consumptions can be explained by operating conditions that were not ideal, e. g. the loss of particles through the cyclone were higher than expected. Both consumed feeds, oxygen and make-up limestone, cause parasitic losses since they consume energy that can not be further utilized. The typical energetic requirements for an ASU is $184\,kWh_{el}$ to provide a ton of oxygen [146]. This leads to an energy penalty by the ASU of 1.7-2.9 %-points. The required heat for first calcination of make-up causes an efficiency penalty of 0.4-2.7 %-points. The sum of the efficiency penalties by oxygen and make-up representing the main share, sums up to 2.6 to 5.3 %-points for steady-state operating points with a carbonator absorption efficiency $> 80\,\%$ (see Table 4.6).

Table 4.6: Range of operating parameters for steady-state operating points with $E_{carb} > 80\,\%$.

Description	Parameter	Unit	Range
Specific heat consumption	Q_s	$MJ/kg_{CO2,capt}$	6.2-8.6
Specific oxygen consumption	$O_{2,s}$	$kg_{O2}/kg_{CO2,capt}$	0.3-0.5
Specific make-up consumption	$CaCO_{3,s}$	$kg_{CaCO3}/t_{CO2,capt}$	28-161
ASU efficiency penalty	$\Delta\eta_{ASU}$	%-pts.	1.7-2.9
Make-up efficiency penalty	$\Delta\eta_0$	%-pts.	0.4-2.7
Sum of efficiency penalties	$\Delta\eta_{CaL}$	%-pts.	2.6-5.3

The obtained experience enables the scale-up and serves as a basis for the definition of favorable operating conditions in terms of high performance and further investigations to be addressed in the next development stage. Moreover, optimization potential for the CaL process is derived from these experimental findings. It can be divided in two groups, the improvement of the sorbent, and the reduction of the heat demand in the calciner.

On the one hand, the CaL process offers improvement potential of the utilized sorbent. The measures mainly cover possibilities to keep the molar conversion, and thus the reactivity as high as possible to counteract the deactivation. There are different options, such as enhanced natural or synthetic sorbents [127]. Natural sorbent enhancement includes additional process steps such as hydration [82, 147] or recarbonization [127, 148] or doping, thermal pre-treatment or chemical treatment [127]. In addition, various possibilities can create synthetic sorbent, e. g. dry mixing and coating or granulation [127]. However, these methods are yet to be investigated. A compromise in the development has to be found between the improvement of sorbent performance and the increase in costs since natural limestone offers the advantage of a widely available and inexpensive material.

On the other hand, the CaL process offers improvement potential in terms of the heat demand in the calciner. The lower the heat ratio between the carbonator and the calciner, the more economically attractive the process is for the retrofit of power industrial plants. An optimized capture unit reduces the input of fuel as well as oxygen, and the size for reactor and air separation unit, and thus investment costs, decrease. To point out the impact, the improvement measures are described as follows and the related potential for heat reduction of 34 % is shown in Table 4.7 for an exemplary operating point.

Table 4.7: Optimization of the calciner energy balance for a steady-state operating point.

Heat flow	Base case	Optimized	Difference	
Heating of circulating sorbent	$298\,kW_{th}$	$111\,kW_{th}$	$-187\,kW_{th}$	-63 %
Flue gas loss	$189\,kW_{th}$	$149\,kW_{th}$	$-40\,kW_{th}$	-21 %
Calcination of sorbent	$108\,kW_{th}$	$108\,kW_{th}$	-	-
Calcination of make-up	$17\,kW_{th}$	$17\,kW_{th}$	-	-
Heat losses	$50\,kW_{th}$	$50\,kW_{th}$	-	-
Total	$662\,kW_{th}$	$435\,kW_{th}$	$-227\,kW_{th}$	-34 %

Heating of the circulating sorbent stream from 600-700 °C calciner inlet to 900-950 °C outlet temperature consumes the highest share of the heat input. In the shown operating point (see Table 4.7), it comprises 45 %. Aside from the temperature difference of between both reactors, the heat demand for the circulating material stream depends on the sorbent properties. Especially, a low CO_2 carrying capacity of the sorbent accompanies a low loading, and consequently leads to significantly higher circulating mass flow or a sorbent looping ratio, respectively. An efficient measure to decrease the heat input is the application of a solid-solid heat exchanger. It transfers the sensible heat of the sorbent leaving the calciner to the one entering. Reducing the temperature difference between both streams to 100°C, the required heat to bring the loaded sorbent at the calcination temperature significantly reduces by 63 %. A possible solution is the application of heat pipes, as shown by Reitz et al. [122].

The **flue gas losses** describe the difference between entering and leaving calciner gas flows. The calciner primary gas consists of oxygen and recirculated flue gas and is heated up to 400-420 °C. This corresponds to the maximum temperatures allowed and available for oxygen-enriched gas preheaters. A higher preheating temperature can not be realized within the pilot plant. The excess heat can be utilized for steam production and power generation in a downstream water/steam cycle. The heat utilized to heat the gas flows of the calciner accounts for 29 % of the fuel input in the shown operating point (see Table 4.7). Thus, the increase of the inlet oxygen concentration offers further reduction potential by reducing the recycled flue gas. This leads to an increase of the combustion temperature while the gas flow rate through the reactor decreases. Due to the endothermic nature of the calcination reaction and the large flow of solids circulating between the reactors, it is possible to operate the calciner with oxygen inlet concentrations of 75 vol.% [128]. Reducing the recirculation gas flow to increase the oxygen concentration in the fluidization gas to 75 vol.% for the given operating point, the heat input is reduced by 21 %. It is an attractive option with the potential to decrease the required heat input as well as the reactor size, and thus the capital costs of the calciner but further long-term tests are required to prove the concept in industrial scale.

Calcination of loaded sorbent and make-up is the essential function of the calciner. The regeneration of the sorbent in this operating point only consumes 18 % of the heat, thereof 15 % for the release of the absorbed CO_2 in the carbonator and 3 % for the first calcination of the make-up. There is only a small improvement potential in terms of reducing the make-up flow.

The **heat losses** are caused by the elevated temperature level of the process compared to the ambient conditions. Technical insulation of the reactor system and its peripheral components reduce the losses. In general, the specific heat losses decrease with increasing reactor size since the volume of the reactor increases disproportionately compared to its surface. In addition, the walls of large scale plants have higher thermal resistances due to an increased wall thickness.

5 Design of a 20 MW$_{th}$ Demonstration Pilot

The preceding chapter shows the results the CaL research in semi-industrial 1 MW$_{th}$ scale. The performance and the long-term operability of the process in terms of effective and efficient CO_2 capture from flue gases of fossil-fired upstream power or industrial plants has been proven. CaL is ready to take the next step towards commercialization and industrial application, respectively. The next step to develop the technology is a scale-up to a demonstration pilot with a thermal equivalent of 20 MW$_{th}$. Therefore, the results of the long-term pilot tests were employed in collaboration with GE Power Sweden (formerly known as GE Carbon Carbon Capture) design and pre-engineer this demonstration pilot. Remaining technological and economic questions can be answered with the help of this kind of a plant.

The technical and economic uncertainties, challenges and risks can be minimized further or even be eliminated by the development to demonstration scale. It allows the implementation of the knowledge gained in 1 MW$_{th}$ scale in the next larger plant size, and the process is verified further. The continuous operation in the demonstration size allows the investigation of long-term effects under real boundary conditions given by the upstream host plant. Further optimisation measures can be implemented with regard to the optimization of heat requirements and excess heat utilization.

The next development stage of the CaL technology aims at verifying further the process under real conditions in a pilot plant and thereby CO_2 from an exhaust gas actually generated in a power plant process. Operating parameters, e. g. reactor temperatures, circulating solids flow, make-up or fuel type, are applied with respect to the results of the 1 MW$_{th}$ pilot plant. Also the investigation of part load capacity under real conditions and circumstances is possible. Particularly with regard to grid loads, which are caused by the feed-in of electricity from wind and solar energy are exposed to strong fluctuations, the future decarbonized power plants can react quickly and flexibly to load changes. This will also affect the downstream CO_2 separation plants, e. g. by the CO_2 concentration of the flue gas stream to be decarbonized.

The design and engineering of the demonstration pilot was carried out in a way that its later operation allows the adaptation and adjustability of the process parameters. It is essential for the demonstration pilot to enable the investigation of CaL technology under real conditions given by a host plant. Thus, a certain degree in flexibility is required to investigate and optimize various parameters. It was decided to focus on the plant design primarily for demonstrating operability. Optimization of heat recovery and heat integration into the host plant, resulting in optimized energetic performance and minimized efficiency penalty, was not selected as main objective of the pilot plant. This approach includes e. g. heat recovery to lower pressure steam systems, no heat recovery from spent sorbent/ash, etc. However, certain components (like gas-gas heater in carbonator flue gas inlet) are not necessarily required for plant operability, but were nonetheless implemented into the pilot plant design to demonstrate feasibility. Consequently, the design with these requirements in mind appears to make more sense than only focusing on energetic efficiency optimization.

Before the 20 MW$_{th}$ demonstration pilot can be built and operated, it must be conceptually designed for the required structural and technical integration in an existing power plant. This chapter describes the design of the demonstration pilot by a detailed process design as well as heat and mass balancing for various operating conditions. These results have already been published [136, 149].

5.1 Boundary Conditions

The conceptual design for a 20 MW$_{th}$ pilot plant was elaborated for the existing 600 MW$_{el}$ power station Emile Huchet Unit 6 as host unit. It is located in Saint-Avold in France and is owned and operated by Uniper [150].

The Emile Huchet Unit 6 power plant is the first 600 MW$_{el}$ installation built with a forced circulation boiler in France. It is fired with bituminous coal and designed to combust low-quality coals with varying particle sizes < 1 mm and high contents of abrasive ash (> 50 %). A particular requirement is to respond rapidly to demand changes, in particular with short start-up times. Inaugurated in October 1981, the power plant has a nominal steam flow of 1,739 t/h with superheater conditions of 181 bar and 543 °C. The reheater pressure is 35 bar. The boiler is designed to provide good combustion efficiency with a minimum pre-processing of the fuel and is built as a tower where all the heat exchanger tubes are placed on top of each other. This arrangement reduces changes in the direction of combustion gases to avoid turbulences. Otherwise, generated turbulences in the gas stream would promote ash-induced abrasion. The gas velocity is thereby reduced from common 18 to 13 m/s. The forced circulation boiler increases the operational flexibility since it provides a faster response to load changes compared to natural or assisted circulation. The firing is realized by 24 burners installed on six levels in each corner. A minimum production of NO$_x$ is realized by tangential arrangement of the firing. Each level is supplied by a vertical bowl mill crushers. Natural gas and heavy oil are used for cold or hot start-up, respectively.

The steam supplied by the boiler is used for power generation by a steam turbine comprising one high pressure, one medium-pressure and two low-pressure sections. Due to the forced circulation mode of the boiler, the steam turbine can operate in sliding-pressure mode. This concept is essentially characterized by a practically constant opening of the inlet turbine valves, thus providing a steam pressure proportional to the power plant throughput. The turbine operation is regularly operated with 40 to 90 % of the nominal load with a valve opening of 90 %. The steam exits the turbines at 50 mbar(a) and 37 °C to be condensed. Therefore, the cooling is provided by a single cooling tower with a capacity of 750 MW.

The flue gas exiting the boiler passes a Selective Catalytic Reaction (SCR) and a Flue Gas Desulphurization (FGD) for reduction of NO$_x$ and SO$_2$, respectively. The SCR unit consists of two reactors. Each is equipped with two catalytic stages and a backup stage. An aqueous solution of ammonia is fed while the flue gas enters with a temperature of 410 °C. The FGD unit applies limestone slurry to absorb the acid components in a single-loop counter current spray tower. The absorber is equipped with three spray levels and a two-stage horizontal mist eliminator. The wet flue gas is sent directly to the stack.

The conditions of the flue gas leaving the host power plant are essential for the CO_2 demonstration pilot. A slip stream of this flue gas is extracted at 48 °C and atmospheric pressure for decarbonization. The load cases of the host power plant thus determine the flue gas composition entering the carbonator of the CaL unit. Under full load operation, the flue gas contains 10.2 vol.% CO_2 whereas the CO_2 concentration is significantly reduced during part load operation. Nevertheless, the equivalent thermal duty per mass of flue gas is also reduced. Consequently, the conditions for CO_2 absorption in the carbonator are less beneficial. The flue gas parameters for the considered part loads for the host power plant are summarized in Table 5.1.

Table 5.1: Flue gas composition of the host plant [150].

Species	Unit	100 %	81 %	66 %	52 %
CO_2	vol.%	10.2	9.5	8.3	7.2
H_2O	vol.%	11.4	11.4	11.4	11.4
O_2	vol.%	6.7	7.4	8.9	10.2
N_2	vol.%	71.8	71.6	71.4	71.2
SO_2	ppmv	55.7	52.2	45.5	39.4
th. Eq.	MJ/kg	1.7	1.6	1.4	1.2

5.2 Plant Setup

The process configuration for the 20 MW_{th} demonstration pilot is based on the 1 MW_{th} pilot plant at Technische Universität Darmstadt and the experience that has been gained from long-term operation [29, 125, 126]. The setup of the demonstration plant is shown in Fig. 5.1.

A slip stream of flue gas is taken from the existing flue gas duct downstream of the FGD of the host power station and is routed to the carbonator reactor via a fan. The flue gas stream is preheated by the off-gas of the carbonator. In the carbonator reactor, the CO_2 is contacted with the CaO in a circulating fluidized bed. The reactor is equipped with a first cyclone to separate the solids from the CO_2 lean flue gas stream. The treated flue gas is sent back to the flue gas duct of the host power station via a heat recovery system and a particulate filter. A part of the solid stream separated by the cyclone is internally recirculated while another part of the loaded sorbent is transferred to the calciner. A loop seal with a cone valve is used to control the circulating solid flow between the reactors. The calciner reactor is built as a circulating fluidized bed reactor where the loaded sorbent is regenerated under oxy-fuel coal combustion. The calcined sorbent is separated by a cyclone and enters a loop seal. One part is internally recirculated in the calciner reactor system and the other part is returned to the carbonator via a sorbent heat recovery system. The solids are cooled in a fluidized bed heat exchanger to moderate the carbonator temperature. Secondary cyclones in both reactor systems limit the solid entrainment to the downstream filters under normal operation and process upsets.

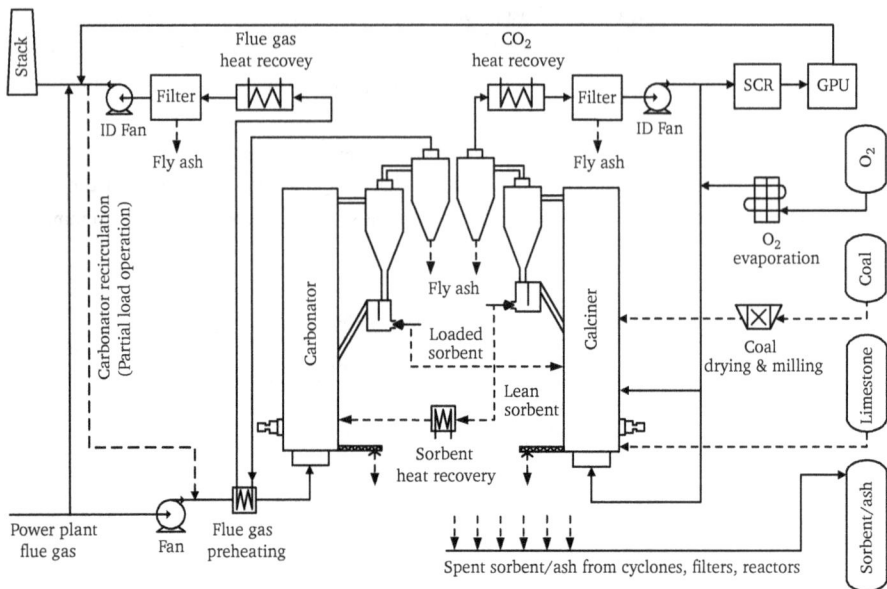

Figure 5.1: Process scheme of the 20 MW$_{th}$ CaL demonstration pilot.

The CO_2 rich flue gas is routed to a Gas Processing Unit (GPU) via a heat recovery system, a particulate filter, and a SCR reducing NO_x emissions. The GPU delivers CO_2 with a purity > 95 mol.% (quality for Enhanced Oil Recovery). A small stream is dried and utilized for fluidization and inertization purposes. Since the pilot unit is solely designed for demonstration of the CO_2 capture process, the CO_2 product is neither utilized nor stored. The CO_2 product stream leaving the GPU is mixed with the treated flue gas stream from the carbonator and is routed back to the stack of the host power station.

Heat recovery from CO_2 lean flue gas, CO_2 rich flue gas, and the sorbent cooling are utilized to produce steam from boiler feed water supplied of the host plant. Since the demonstration pilot is designed for flexible operation rather than efficient power production, the excess heat removal is designed with lean auxiliary system without optimized heat utilization. The steam is used either inside the plant for coal drying or fluidization of the loop seals or is exported to the infrastructure of the host power station. The coal for the demonstration pilot is supplied by the coal yard of the host power plant. In the plant, the fuel is dried and milled to the targeted particle size of approx. 80 μm (d_{50}) before being sent to the calciner. Limestone is continuously supplied to the calciner system and stored in a hopper. Liquefied oxygen is supplied by tanks and is evaporated before mixing with calciner recirculation gas. The spent sorbent and ash separated by the filters and cyclones is cooled and temporarily stored before being disposed.

Compared to the experimental setup in the 1 MW$_{th}$ pilot plant, the concept for the 20 MW$_{th}$ demonstration pilot is slightly modified. Instead of applying cooling lances to moderate the carbonator temperature under the exothermic reaction, the solids entering the carbonator are cooled by a fluidized bed heat exchanger. Cooling down the entering solids allows a more homogeneous temperature profile in the carbonator loop. The particles in the loop seal have a higher temperature, thus reducing the thermal power required in the calciner. A further difference is the coupling concept between the reactors. Instead of a screw conveyor to transport the solids from carbonator to calciner, a loop seal with a cone valve is used. The flow vice versa is also controlled by a loop seal with a cone valve. The feasibility of using a cone valve for solid flow control was successfully tested in the $1\,MW_{th}$ pilot in two out of six test campaigns (see Sec. 4.3.1). This concept was chosen for the demonstration pilot since a scaled-up screw conveyor bears a higher risk for downtimes and increased maintenance efforts. A screw conveyor needs an additional cooling system and rotating parts with packing glands that are susceptible to upsets or leakage of solids. Another difference is the pre-heating of fluidization gases to both reactors. The flue gas led to the carbonator is heated by the CO_2 depleted off-gas of the reactor instead of electrical pre-heating. The calciner fluidization gas is not pre-heated because the temperature level of the recirculation gas after compression is already $220\,°C$ and an appropriate heat source is not available.

5.3 Definition of Operating Parameters

The design of the reactor system and the auxiliary components is related to the process setup. Therefore, a process model of the $20\,MW_{th}$ demonstration plant was created [149] and applied to establish mass and heat balances considering the experimental results in $1\,MW_{th}$ scale and the boundary conditions given by the host plant. The applied process model uses flowsheet simulation in the ASPEN PLUS™ V8.8 simulation environment and was initially described by Ströhle et. al. [124]. The process model includes a 1D reactor model with hydrodynamics based on a FORTRAN™ interface to include dedicated mathematical formulations. The carbonation as well as the sulphation reaction are implemented. The core components are two intercoupled CFB reactors. This model was further enhanced and validated with experimental data from pilot tests [151], so that a reliable basis for the reactor design and the evaluation of demonstration plant performance could be derived with its help.

5.3.1 Design Parameters

The main goal of the demonstration pilot is the demonstration of CaL applicability in $20\,MW_{th}$ scale. The thermal power equivalent of the flue gas fed from the upstream host plant and the thermal power input of the calciner are in sum $20\,MW_{th}$. The total CO_2 capture rate of the demonstration pilot should be at least $90\,\%$ including the amount of CO_2 absorbed from the slip stream of the flue gas from the power plant, and the CO_2 released during oxy-combustion in the calciner. The CO_2 absorption efficiency of the carbonator should be approx. $80\,\%$. Furthermore, the demonstration pilot will provide first insights with respect to thermal integration of the CaL process into existing power plant structures. The demonstration pilot design parameters and its range is shown in Table 5.2

Table 5.2: CaL design parameters for the demonstration pilot.

Description	Parameter	Unit	Design	Range
Carbonator average temperature	T_{carb}	°C	650	650-675
Carbonator inventory	$W_{s,carb}$	kg/m^2	1,000	1,000-1,500
Carbonator superficial gas velocity	$u_{0,carb}$	m/s	5.0	3.0-5.5
Carbonator vol. CO_2 inlet fraction	$v_{CO2,carb,in}$	vol.%	10.2	7.2-10.2
Carbonator absorption efficiency	E_{carb}	%	80	78-82
Make-up ratio	MUR	mole$_{Ca}$ /mole$_{CO2}$	0.38	0.36-0.48
Sorbent looping ratio	LR	mole$_{Ca}$/mole$_{CO2}$	13	9-17
Calciner average temperature	T_{calc}	°C	950	900-950
Calciner inventory	$W_{s,calc}$	kg/m^2	500	-
Calciner superficial gas velocity	$u_{0,calc}$	m/s	5.0	2.5-5.1
Calciner heat input	$\dot{Q}_{th,calc}$	MW$_{th}$	12.2	6.7-12.2
Calciner vol. O_2 inlet fraction	$v_{O2,calc,in}$	vol.%	40	40-60
Calciner vol. O_2 outlet fraction	$v_{O2,calc,out}$	vol.%$_{db}$	2.0	0.7-2.4
Total CO_2 capture efficiency	E_{tot}	%	>90	>90
Thermal capacity	$\dot{Q}_{tot}$	MW$_{th}$	20	10-20

The sorbent chosen for the demonstration pilot is the Messinghausen (MF) with a d_{50} of 180 μm. After proving its usability in terms of of CO_2 absorption and regeneration as well as its mechanical stability in all 1 MW$_{th}$ tests, this limestone was chosen. Under a wide range of fluidization velocities between 2 and 6 m/s in carbonator or calciner, respectively, conditions concerning sufficient fluidization and entrainment was were fulfilled.

The design temperatures of carbonator and calciner are 650 °C and 950 °C, respectively. As shown during pilot operation, this carbonator temperature appears to be the optimum for CO_2 absorption. In terms of sorbent regeneration, the pilot plant results show that at least 925 °C allow a full calcination of the sorbent. The conservative design decision for 950 °C was made to ensure a sufficient difference to equilibrium conditions and thus to avoid any limitations to achieve full calcination of the circulating sorbent. In case of dilution with recirculation gas to sufficiently lower the CO_2 partial pressure in the calciner, it is envisaged to test lower calcinator temperatures to decrease the heat demand.

The main operating parameters of the carbonator (specific inventory, make-up and sorbent looping ratio) with respect to the CO_2 absorption efficiency were determined iteratively with the help of the findings in 1 MW$_{th}$ scale and the validated process model. To define the basis of parameters, the active space time of the carbonator (see Sec. 4.4.3) was used. As indicated, a carbonator absorption efficiency of 80 % was achieved at $\tau_{active,carb} = 50$ s under the given conditions during 1 MW$_{th}$ pilot tests ($\nu_{CO2,carb,eq} = 1.15$ vol.%, $\nu_{CO2,carb,in} = 13$ vol.%, $k_s\varphi = 0.31$ 1/s). The conditions of the demonstration plant had to be changed due to the different boundary conditions. The inlet CO_2 concentration $\nu_{CO2,carb,in}$ of 10.2 vol.% is lower, whereas the equilibrium ($\nu_{CO2,carb,eq}$ of 1.0 vol.%) is almost equal. As the limestone and fluidization regime are intended to be the same, the apparent reaction rate constant $k_s\varphi$ was unchanged. Due to the changed conditions, the active space time required to achieve an carbonator efficiency of 80 % increased to 62 s since the conditions in the reactor demonstration pilot are less beneficial for the absorption reaction. Parameter values from 1 MW_{th} pilot tests were identified to achieve the required active space time and acted as the basis for the scale-up. Consequentially a specific inventory of 770 kg/m^2 with a make-up ratio of 0.18 mole$_{Ca}$/mole$_{CO2}$ and a high sorbent looping ratio of 18 mole$_{Ca}$/mole$_{CO2}$ were derived from the evaluation of pilot operation. The molar Ca conversion of the sorbent τ_{carb} was 0.06 mole$_{CaCO3}$/mole$_{Ca}$.

Based on these values, an iterative definition of the design operation case was made for the 20 MW_{th} demonstration pilot. Thereby, the carbonator inventory was fixed to 1,000 kg/m^2. This is a conservative assumption to guarantee a sufficient inventory of solids in order to achieve the maximum carbonation conversion. Then, the parameters of make-up and sorbent looping ratio were iterated. The iteration considered an increase of make-up while decreasing the sorbent looping ratio under the premise of a carbonator efficiency of 80 %. The decision to decrease the sorbent looping ratio was based on the consideration that a higher make-up and a lower sorbent looping ratio reduce the heat requirements in the calciner more than vice versa. At the end of the iteration process, a make-up ratio of 0.38 mole$_{Ca}$/mole$_{CO2}$ and a sorbent looping ratio of 13 mole$_{Ca}$/mole$_{CO2}$ were defined. All parameters led to an active space time of $\tau_{active,carb} = 59$ s, as given scale-up basis derived from pilot testing, and a calculated carbonator efficiency of 81.3 %. The molar Ca conversion X_{carb} in the design case of 0.06 mole$_{CaCO3}$/mole$_{Ca}$ is comparable with the experimental results.

The calciner inventory was defined accordingly. During 1 MW_{th} pilot operation, a calciner active space time (see Sec. 4.4.3) of at least $\tau_{active,calc} = 10$ min enabled calcination efficiencies close to 100 %. The calciner inventory was up to 350 kg/m^2 with residence times in the reactor of 30-50 s while the molar Ca conversion of the circulating sorbent stream τ_{carb} was 0.06 mole$_{Ca}$/mole$_{CO2}$. The specific inventory of the calciner was chosen to be 500 kg/m^2 in the design case. The corresponding active space time of the demonstration pilot was calculated by the inventory, the circulating sorbent stream and its carbonate content. The value of 23 min exceeds the requirements of 10 min derived from pilot operation significantly. This conservative approach was applied since the inventory was comparably low during 1 MW_{th} pilot operation as a consequence of the reactor geometry. It is expected that an increased inventory provides better combustion behaviour. In addition, the operation with very high oxygen concentrations is only possible with a high solid inventory in the calciner.

Both CFB reactors are designed with a superficial gas velocity of 5 m/s. A sufficient fluidization is provided to achieve the fast fluidized bed regime in both the carbonator and the calciner. Thus, a good mixing of gas and solids as well as a high entrainment is ensured. The high superficial gas velocities also allow a significant reduction for part-load operation down to 2.5 m/s. These specifications are based upon the operational experience in $1\,MW_{th}$ scale. The carbonator was fluidized in the range of 2.2-3.0 m/s while the calcier ran at 4.5-6.0 m/s due to the given limitations of the pilot. In all the cases, sufficient fluidization and entrainment was guaranteed. The utilized sorbent allows this range of fluidization velocity. In addition, the part-load behaviour was calculated and proven by Haaf et. al. [149].

The demonstration pilot is designed to handle oxygen concentration in the calciner primary gas up to 60 vol.% in order to examine the influences on the process in terms of efficiency improvements. In the design case, the oxygen concentration is set to 50 vol.%, as operated in the pilot plant. The calciner outlet oxygen concentration is set to $2\,vol.\%_{db}$. The oxygen excess is chosen as small as possible but high enough to guarantee the burnout of the solid fuel. Lower oxygen excess leads to insufficient combustion behaviour. A higher oxygen excess is not desired in terms of an increased power demand by an ASU.

The dimension of the CFB reactor systems are specified based on the superficial gas velocities. Both CFB reactors are rectangular. The carbonator cross-sectional area is $2.25\,m^2$ with the dimension of 1.5 m x 1.5 m whereas the calciner geometry is 1.48 m x 1.48 m with a cross-sectional area of $2.2\,m^2$. The reactor height of both carbonator and calciner is 20 m. The height to diameter ratio of 11 is close to common values of CFB system for Geldart particles of group B [101]. Furthermore, the reactor heights are driven by layout requirements. Gravitationally driven flow for solids through cyclones and loop seals is ensured by adequate slope of piping. The achievement of the system performance in the design case with the given reactor geometry was confirmed with validated 3D CFD models [152–155].

The $1\,MW_{th}$ scale pilot tests showed the crucial impact of coal composition and particle size (see Secs. 4.5.3 and 4.5.4). The coal type and preparation influences the accumulation of impurities in the circulating sorbent stream such as ash and calcium sulphate. Thus, different coal types are considered for the demonstration plant. The coals available at the host site were screened and two different types of coal were selected to be utilized in the demonstration pilot (see Table 5.3). One of the is El Cerrejon (EC) with a low sulphur and a high ash content, the other coal is US High Sulphur (UHS) with

Table 5.3: Composition of fuels (as received) for the demonstration pilot.

Species	Unit	El Cerrejon (EC)	US high sulphur (UHS)
Carbon	wt.%	64.68	71.42
Hydrogen	wt.%	5.19	4.94
Nitrogen	wt.%	1.20	1.52
Oxygen	wt.%	5.08	5.67
Sulphur	wt.%	0.84	2.02
Moisture	wt.%	9.48	7.00
Ash	wt.%	13.53	7.35
LHV	MJ/kg	25.43	28.87

an elevated sulphur content of 2.02 wt.% and a low ash content. The design particle size given by the mill is expected to be 55 wt.% < 90 μm with a d_{50} of approx. 80 μm. The coal particle size can be varied by changing the setting in the milling system to enable the investigation of the fuel particle size.

5.3.2 Operating Points

Based on the long-term pilot tests in $1\,MW_{th}$ scale, in total eleven operating points were defined. Each Operating Point (OP) represents an alteration of a crucial process parameter, except of OP11. This operating point is optimized with respect to the CO_2 capture performance and heat requirements. The design load case and the results derived from the $1\,MW_{th}$ pilot tests led to the following operating points (see Table 5.4). It is worth mentioning that a further variation of the parameters will be possible within the boundary conditions given by the total system. A description of each operating point is given:

- OP01: Full load operation case
  This load case represents the design conditions in the full load scenario of the demonstration plant. All equipment and subsystem design is derived from the boundary conditions of this operating point. As described before, the carbonator temperature is 650 °C and the calciner temperature 950 °C, respectively to ensure the complete calcination of the circulating sorbent. The calciner fuel is Colombian hard coal.

- OP02: High carbonator temperature
  For this OP, the carbonator temperature is increased to 675 °C. The volumetric equilibrium concentration here increases from 1.2 to 2.1 vol.% and the carbonation reaction runs faster. Also, an increased carbonator temperature reduces the heat demand in the calciner.

- OP03: Low calciner temperature
  This OP lowers the calciner temperature to 900 °C to investigate whether the temperature is sufficient to ensure complete regeneration of the loaded sorbent. In the design case, 950 °C are considered to avoid any limitations in terms of kinetic and diffusion controlled issues. The lower calciner temperature implies a lower heat demand in the calciner to bring the circulating solid stream and the recirculation gas stream. Less fuel is burnt and consequently less oxygen is required for operation.

- OP04: High make-up feeding rate
  The influence of the make-up feeding is investigated by an increased make-up ratio in OP4. The MUR is increased from 0.38 to 0.48 $mole_{Ca}/mole_{CO2}$ in expectation of a higher sorbent activity compared to the design case. However, an increased heat and oxygen demand is expected with increasing make-up feed, because more fresh limestone needs to be calcined.

Table 5.4: Boundary conditions of the nominal operating points.

Parameter	Unit	01	02	03	04	05	06	07	08	09	10	11
Load case host plant	%				100				81	66	52	100
Load case pilot plant	%					100					50	100
Type of coal	-			EC				UHS		EC		
Carbonator temp.	°C	650	675					650				
Calciner temp.	°C	950		900				950				900
Make-up ratio	Ca/CO_2		0.38		0.48				0.38			
Calciner O_2 inlet conc.	vol.%			40		50			40			60
Carbonator inventory	t/m^2			1.0			1.5		1.0			1.5

- OP05: Elevated oxygen concentration in calciner primary gas
  The oxygen concentration at the calciner inlet is increased from 40 to 50 vol.%. This number is in accordance with the setting in the 1 MW_{th} pilot plant. An increased oxygen concentration lowers the recirculation rate of the primary gas. Hence, less gas needs to be heated and coal and oxygen are saved. However, the increase of oxygen concentration is limited in order to avoid local hot spots causing sintering or partial melting of the particles.

- OP06: High specific carbonator inventory
  The specific carbonator inventory in the design case of 1,000 kg/m^2 is increased to 1,500 kg/m^2 to investigate the influence of particle residence time in the carbonator. It is expected to increase the loading of the sorbent by an increased residence time under the condition of a sufficiently active sorbent.

- OP07: High sulphur coal
  In order to show the impact of a different coal type on pilot operation, the 'US high sulphur' coal is used instead of the reference coal 'El Cerrejon'. The main differences between these two types of coal are the higher sulphur content and a slightly increased lower heating value. Especially the sulphur content facilitates deactivation of the sorbent due to $CaSO_4$ formation and, thus to an increased share of inert material circulated between the reactors.

- OP08: Reduced load of the host plant (81 %)
  The reduced load of the host plant (81 %) is considered in this case. Reduced load in the upstream plant leads to a changed composition of the flue gas entering the carbonator (see Table 5.1). The CO_2 concentration is lowered from 10.18 to 9.54 vol.% as well as the thermal equivalent of the flue gas from 1.75 to 1.43 MJ/kg.

- OP09: Reduced load of the host plant (66 %)

  An even further reduced load of the host plant (66 %) is considered in this case. The CO_2 concentration of the flue gas entering the carbonator (see Table 5.1) is decreased from 10.18 to 8.33 vol.% as well as the thermal equivalent of the flue gas from 1.75 to 1.43 MJ/kg.

- OP10: Reduced load of the host plant (52 %) and the demonstration unit (50 %)

  This operation point represents the minimum load of the demonstration pilot (50 %). It is linked to the minimum load of the upstream host plant (52 %). The CO_2 concentration of the flue gas to be decarbonized is 7.21 vol.% corresponding to a thermal equivalent of 1.25 MJ/kg.

- OP11: Optimized capture efficiency operation

  In contrast to the described operating points above, OP11 considers an alteration of more than a single parameter. The boundary conditions are adapted to optimize conditions for CO_2 absorption. The thermal load of the calciner is decreased by lowering the calciner temperature to 900 °C and increasing the oxygen inlet concentration to 60 vol.%.

5.3.3 Heat and Mass Balance

The balance of mass and heat flows is essential to the design and engineering of all subsystems and auxiliaries. The specifications of the demonstration pilot equipment is defined by these values. Limitations by the system boundaries need to be considered. In general, 10 % overdesign of the design case are allowed in terms of mass flows and cooler duties. The mass and heat balance for the derived design case is described as follows, and the range of the specifications is given.

In the design case, a flue gas slip stream with a thermal equivalent of 7.9 MW_{th} is fed to the carbonator. The calciner requires a heat input of 12.2 MW_{th}. The global mass flows of feeds entering and products leaving the carbonator and calciner system, respectively, as well as the heat streams are shown in Table 5.5 for the design case (OP1) and the range derived from the other operating points (OP02-11).

A flue gas flow of 16,200 kg/h is fed to the carbonator for decarbonization in the design case. The CO_2 depleted flue gas flow of 14,100 kg/h leaves the reactor whereat 81.3 % of the CO_2 is absorbed by the circulating solid. The sorbent circulation is 46,500 kg/h of loaded sorbent from carbonator to calciner and 45,100 kg/h of lean sorbent vice versa corresponding to the design sorbent looping ratio of 13 $mole_{Ca}/mole_{CO2}$. To heat up the circulating sorbent and to run the calcination, 1,800 kg/h of coal are burned with 3,600 kg/h oxygen in the calciner with an oxygen excess of 2 vol.%$_{db}$. The required make-up feed for a make-up ratio of 0.38 $mole_{Ca}/mole_{CO2}$ is 2,200 kg/h. This mass flow of fresh limestone is added to achieve the required activity of the material. A CO_2 product mass flow of 14.400 kg/h leaves the calciner whereof 45 % or 6,400 kg/h are used for flue gas recirculation to dilute the oxygen and moderate the reactor temperature. The rest is sent to the GPU. A water mass flow of 900 kg/h is condensed and a CO_2 product stream of 7,200 kg/h is sent to the stack. To keep the solid inventory of the total system constant, a stream of 1,600 kg/h spent sorbent is extracted. The total balance of feeds and products entering and leaving the CaL system boundaries in the design case is 23,800 kg/h.

Table 5.5: Range of the main mass and heat flows of the demonstration pilot.

Parameter	Unit	Design	Range
Flue gas flow to carbonator	t/h	16.2	10.0-17.5
Coal flow to calciner	t/h	1.8	1.0-1.8
Make-up sorbent flow	t/h	2.3	1.9-3.0
Sorbent circulation	t/h	46.5	22.7-48.9
Oxygen flow to calciner	t/h	3.6	2.0-3.8
CO_2 product	t/h	7.2	3.9-7.3
Condensate	t/h	0.9	0.5-1.0
Calciner heat input	MW_{th}	12.2	6.7-12.2
Sorbent cooling	MW_{th}	4.4	1.8-4.4
Carbonator flue gas cooling	MW_{th}	0.8	0.6-0.9
Calciner flue gas cooling	MW_{th}	4.1	2.2-4.1

The heat from the CO_2-lean flue gas, the CO_2 rich flue gas and the sorbent heat recovery system is utilized to produce steam from boiler feed water. This is a major benefit of the CaL technology since the produced steam can either be used as a heat source in the demonstration pilot or exported to the host plant. The heat of the CO_2 depleted flue gas leaving the carbonator is not fully used for steam generation. Before entering the heat exchanger, the flue gas heats the carbonator primary gas from 81 to 350 °C while it is cooled down from 650 to 360 °C. Then, 0.8 MW_{th} is transferred to the water/steam cycle to cool down the gas to 180 °C. The extracted heat from the calciner off-gas is significantly higher. The heat exchanger extracts 4.1 MW_{th} to cool down the gas from 950 to 180 °C. The calciner primary gas is not heated since it leaves the fan with 220 °C after compression. Another heat recovery takes place from the sorbent leaving the calciner and entering the carbonator. The circulating particles are cooled down from 950 to 590 °C to control the carbonation temperature. Here, a heat stream of 4.4 MW_{th} is extracted for steam production. As shown, especially the amounts of heat extracted from calciner off-gas and the sorbent heat recovery are significant at a high temperature level. Thus, valuable sources for steam and subsequent power production are offered for full-scale application with complex heat integration.

5.4 Assessment of the Demonstration Plant

The evaluation of the demonstration pilot is carried out taking into account the aspects of total CO_2 capture efficiency, the thermal duty, heat ratio, the specific heat and oxygen consumption as well as the sorbent composition.

The carbonator absorption efficiency and the total CO_2 capture efficiency is of great importance for the commercialization of the technology. The demonstration pilot is intended to operate with a carbonator absorption efficiency of 80 % with a total CO_2 capture efficiency of at least 90 % for the planned investigations. The desired carbonator efficiency is reached for all operating points except of OPs 2, 8 and 9. There it is slightly below with 77 %. The lower CO_2 absorption efficiency in the carbonator can be explained by the prevailing conditions in each case, e. g. the carbonator temperature or the CO_2 flue gas concentration at the carbonator inlet. However, the total CO_2 capture efficiency for all cases is above 90 % for all cases.

The total thermal duty as well as the breakdown for carbonator and calciner reactor systems are shown in Fig. 5.2. In addition, the heat ratio is depicted to show relation between both reactors. As shown, all operation points are close to full load of $20\,MW_{th}$. OP10 is an exception since it represents the part load of the pilot. Other operating points with a thermal duty slightly below the full load imply restrictions to stay within design limitations. The heat ratio in the design case is 1.58. Parameter variation in OP3 to OP5 slightly decrease the heat ratio to around 1.35. A further reduction to 1.0 is reached in OP11. In this case, all aforementioned parameters are applied for optimization in that case. It is also obvious that the part load conditions of the host plant in OP8 to OP10 strongly influence the heat ratio. For this operating points 8 and 10, the decreasing CO_2 concentration in the carbonator strongly influences the process performance in terms of the required heat demand. A heat ratio up to 1.92 is shown. The influence of the fuel type is shown with OP7. The applied US high sulphur coal leads to an increased heat ratio up to 1.98. This is mainly caused by an increased inert share of material in the circulating solid stream that has to be heated up in the calciner.

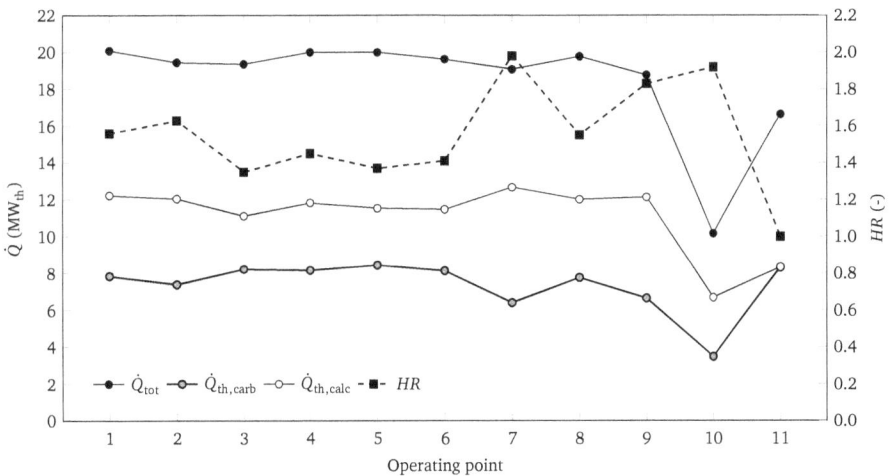

Figure 5.2: Thermal load of each reactor system and corresponding heat ratio.

Fig. 5.3 shows the specific oxygen consumption and the molar Ca conversion for each operating point. The specific oxygen demand expresses the amount of oxygen fed per ton of CO_2 captured. Two effects decrease the oxygen demand for the process. On the one hand, the reduction of the calciner temperature (OP3) and the increase of the oxygen inlet concentration (OP5) reduces the required fuel, and thus the oxygen input. On the other hand, the molar Ca conversion of the solid influences the plant performance. Operation with a higher make-up ratio (OP4) and a higher carbonator inventory (OP6) results in an increased molar Ca conversion sorbent. Thus, the consumption of fuel and oxygen decreases since less material needs to be circulated. Especially OP11 yields to a rather low specific oxygen demand of $0.4\,\mathrm{kg_{O2}/kg_{CO2,capt}}$. The combination of the mentioned measures increases the carbonate conversion to 9%. The increased carbonator inventory and the reduction of the calciner temperature lead to this result while the oxygen inlet concentration is elevated at the same time. In contrast, the lowest carbonate conversion of OP9 leads to a rather high heat and oxygen demand. The low carbonate content is mainly due to the low CO_2 concentration of the flue gas to be decarbonized. This is a result of the part load operation of the upstream host plant.

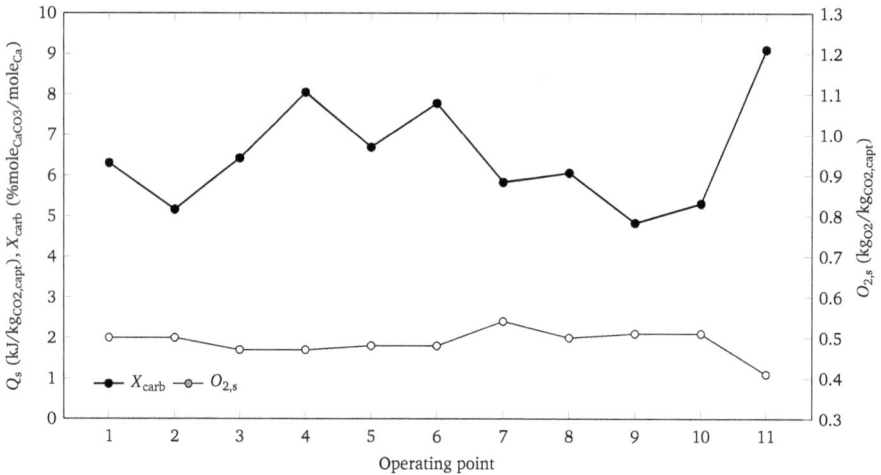

Figure 5.3: Specific heat and oxygen consumption and carbonation conversion.

The composition of the circulating solid stream is shown in Fig. 5.4. The graph shows the fractions of $CaCO_3$, $CaSO_4$ and ash for the carbonator outlet. This stream is transferred to the calciner and carries the CO_2 absorbed in the carbonator reactor for all operating points. With respect to the absorption efficiency of the carbonator, the $CaCO_3$ fraction is important. As it can be seen by comparison with molar Ca conversion depicted in Fig.5.3, high $CaCO_3$ fractions are achieved. The $CaCO_3$ is in a range of 7.8-14.7 wt.% corresponding to a molar Ca conversion of 0.048 to $0.09\,\mathrm{mole_{CO2}/mole_{Ca}}$. In the design case, 10.3 wt.% of $CaCO_3$ is formed in the circulating solid stream. The corresponding molar carbonate content is $0.06\,\mathrm{mole_{CO2}/mole_{Ca}}$. This number is further increased to 14.7 wt.% or $0.09\,\mathrm{mole_{CO2}/mole_{Ca}}$,

5 Design of a 20 MW$_{th}$ Demonstration Pilot

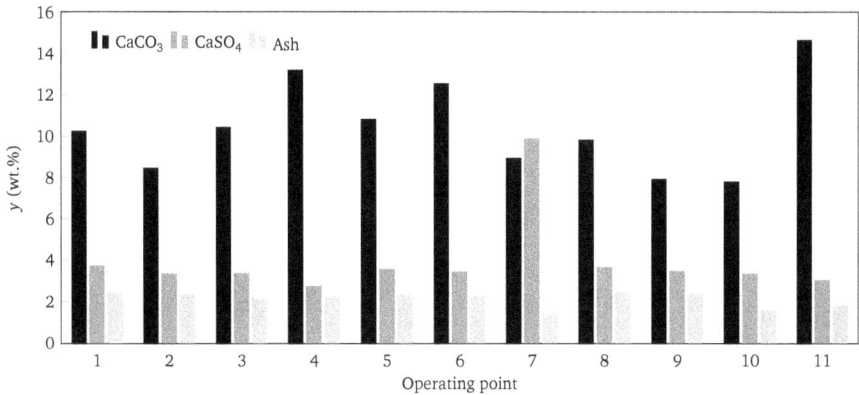

Figure 5.4: Sorbent composition in the solid stream leaving the carbonator.

respectively in OP11. The $CaCO_3$ fraction is mainly increased by a higher make-up ratio (OP4) and a higher carbonator inventory (OP6). The $CaSO_4$ fraction is rather constant around 3.5 wt.% for all operating points except for OP4 and OP7. The increased make-up feed in OP4 reduces this fraction to 2.3 wt.%. In contrast, the increased sulphur content of the coal in OP7 results in high inert $CaSO_4$ fraction of approx. 10 wt.%. For all operating points, the low ash content of around 1.4-2.5 wt.% becomes apparent. The poor separation efficiency of the cyclone for small sized ash particles is the reason for this effect.

6 Conclusions and Outlook

The CO_2 capture from power or industrial plants is an important contribution to meet the ambitious climate goals of the Paris Agreement. Carbonate Looping is therefor a promising second generation CCS technology that is characterized by significant lower efficiency penalties and CO_2 avoidance costs compared to other technologies.

In the context of this work, the Carbonate Looping process was investigated with long-term pilot tests in semi-industrial $1\,MW_{th}$ scale. The pilot plant consisting of two interconnected circulating fluidized bed reactors was upgraded to enhance its operability and to enable realistic operating conditions for later commercial application. The upgrading included coupling concept to interconnect calciner and carbonator based upon a loop seal equipped with a cone valve. This kind of concept is common practice in industrial CFB reactors. Realistic operating conditions in terms of coal-originated flue gas for decarbonization and oxyfuel combustion in the calciner were considered in the plant setup. Additionally, the fuel flexibility of the system was enhanced to utilize different types of fuel. Thus, the flexibility of the process in terms of fuel type (hard coal and lignite) and particle size (pulverized up to coarse) was considered. Based on these prerequisites, six long-term tests campaigns with different fuels and limestone sorbents were carried out to address the flexible and reliable operation of the CaL process under a wide range of operating parameters. The obtained results provide the valid basis for scaling-up the process.

The directly heated CaL process was investigated in $1\,MW_{th}$ scale for more than 2,000 h operation with intercoupled CFB reactors. Thereof, 1,500 h showed CO_2 capture by the CaL process, i.e. a stable CO_2 absorption in the carbonator and CO_2 desorption in the calciner, respectively. Thus, pilot operation proved the continuous and efficient CO_2 capture and the operational stability of the process realizing CO_2 absorption rates in the carbonator higher than 90 % and overall CO_2 capture rates higher than 95 %. Extensive analysis of samples taken during operation enabled the evaluation of sorbent properties, such as reactivity, composition and change of particle size distribution. All in all, the measured data and sorbent analysis allowed extensive heat and mass balancing to determine the important process parameters.

Hydrodynamic stability as an essential factor for continuous and steady process operation was proven. For this reason, different coupling concepts were utilized with respect to the application in a demonstration plant. The calciner to carbonator coupling was tested with two different concepts: a connection with two loop seals and a loop seal equipped with a cone valve. Pressure sealing and continuous solid circulation between the reactors were realized and both coupling variants showed stable hydrodynamics, and thus its application is feasible in larger scale.

Experimental results from the pilot tests show a high CO_2 capture efficiency $> 90\,\%$ for a wide range of conditions. The tests were carried out with two different sorbents and various fuel types varying with particle sizes. Conventional and commercially available fuels and sorbents are suitable to achieve a carbonator absorption efficiency of more than 80 %. The evaluation of steady-state operating points included the influencing factors on the CO_2 absorption efficiency of the carbonator such as reactor temperature and inventory as well as sorbent looping and make-up ratio. All parameters were summarized and considered in the active space time expression. It was pointed out that an active space time $> 50\,s$ leads to carbonator absorption efficiencies $> 80\,\%$ with a corresponding molar Ca conversion of 0.04-$0.06\,mole_{CaCO3}/mole_{Ca}$. An operating window with a reactor temperature of 650-$675\,°C$, an inventory $> 800\,kg/m^2$, a make-up ratio $> 0.15\,mole_{Ca}/mole_{CO2}$ and a sorbent looping ratio of 10-$15\,mole_{Ca}/mole_{CO2}$ was identified as the basis for process scale-up.

Furthermore, the impacts of oxy-calcination conditions with CO_2 concentrations up to $80\,vol.\%$ on the sorbent regeneration efficiency were assessed. To achieve highly efficiency sorbent calcination, calciner temperatures of at least $925\,°C$ are beneficial. For the achieved sorbent conversion of 0.04-$0.06\,mole_{CaCO3}/mole_{Ca}$, an active space time $> 10\,min$ leads to calcination efficiencies $> 90\,\%$. Additionally, the influence of the calciner fuel was investigated. The process proved its operability with various hard coals and lignite in different particle sizes. Depending on fuel type and particle size, the accumulation of impurity fractions such as calcium sulphate and ash between 4 and $20\,wt.\%$ was observed.

Among other things, the required heat input to capture CO_2 with the CaL process and the related efficiency penalties were evaluated. The specific heat input for efficient capture was determined with 6.2-$8.6\,MJ$ per kg of CO_2 captured. The main driver of the calciner fuel input is the heating of the circulating sorbent to calcination temperatures. In addition, the efficiency penalties caused by the application of the CaL process were determined. Make-up and ASU together lead to an efficiency drop of an upstream power plant of 2.6-$5.3\,\%$-points.

The experimental results obtained by this work are a milestone in the development process of the CaL technology. Long-term demonstration of the process in $1\,MW_{th}$ scale provides the reliable basis for further research in the next development stage. Therefore, a demonstration plant with a thermal size of $20\,MW_{th}$ was conceptualized. The design includes a detailed plant setup based on the operational experience in the $1\,MW_{th}$ pilot, e. g. the application of loop seals equipped with cone valves for reactor coupling. Based on the setup of the demonstration pilot and with the help of a validated process model, the design parameters were defined and verified. The demonstration pilot is also designed for various operating points to address and validate the results from semi-industrial pilot testing. The range of operating points comprises the reactor temperatures of calciner and carbonator, the make-up ratio, the carbonator inventory, the load of the upstream host plant, the oxygen inlet concentration and the fuel type of the calciner as well as part load conditions for the capture unit. The heat and mass balancing for all operating cases defined the boundary conditions for the engineering of the demonstration unit. Furthermore, the assessment of the demonstration unit showed the expected aspects of total CO_2 capture efficiency, the thermal duty, heat ratio, the specific heat and oxygen consumption as well as the sorbent composition.

As an ensuing result of this work, the design of the $20\,MW_{th}$ demonstration pilot served for the assessment of engineering, operational, logistical, safety or permitting aspects by GE Power Sweden (formerly known as GE Carbon Capture). The engineering data include the process scheme, type and material of process equipment, required process instrumentation, including analytical equipment and control loops as well as required sizes and materials for major piping or ducts. In addition, a the plant layout was created to arrange the different subsystems, equipment, and interconnecting piping of the pilot plant. All the acquired information could be used to estimate the costs for this next development step. The total cost including contingency (without risk & profit) for five years of operation are expected to be 78.1 million€, thereof 63.6 million€ for investment and 14.5 million€ for operational costs [136]. The results act as a basis for investments into a larger-scale 20 MW$_{th}$ demonstration pilot. Since there are no insuperable obstacles, it facilitates the design of a large-scale demonstration unit with the aim of commercializing the technology in the near future.

The CaL technology is very interesting for the application at industrial plants, such as cement or steel, waste incineration or biomass power plants. These kind of plants require a scaling factor of 3-5 of the 20 MW$_{th}$ demonstration pilot. The existing host plant infrastructure and other process equipment can be used to minimize costs. Especially the fuel flexibility of the CFB calciner allows the utilization at very different host processes. The CaL process provides the possibility to combust local coals, biomasses or alternative fuels such as refuse-derived fuel. Noteworthy to mention as an additional advantage are the possibilities of biomass co-combustion and waste utilization in the calciner. Especially the cement manufacturing process offers great potential for the CaL technology. By utilizing the synergy in terms of $CaCO_3$ as feedstock for both processes, significant savings in CO_2 emissions can be realized. Thereby, the CaO-rich purge from the calciner of the CaL unit replaces significant fractions of the initial feedstock of the cement process. Therefore, required adaptions for the CaL process, such as higher CO_2 concentration in the carbonator, smaller particle sizes and a more active sorbent by high make-up rates, are currently being investigated.

Further research is needed to address improvement in terms of heat requirements of the process. Heat transfer between the circulation sorbents streams offers a great improvement potential and can be realized with heat pipes to significantly reduce the fuel input to the calciner. Furthermore, it is possible to provide indirectly the calcination heat with the aim to omit the air separation unit. The indirectly heated CaL process offers a purer CO_2 product and a lower efficiency penalty. However, these improvements still need to be addressed in extensive pilot tests in various scales.

Nomenclature

Latin symbols

A	m^2	Area
$CaCO_{3,s}$	$kg_{CaCO3}/t_{CO2,capt}$	Specific limestone consumption
d_{50}	μm	Median diameter
d_p	μm	Particle diameter
d_p^*	-	Dimensionless particle diameter
E_{carb}	%	Carbonator absorption efficiency
E_{calc}	%	Calciner efficiency
E_{tot}	%	Process capture efficiency
F	$mole/m^2s$	Molar flow
f_{active}	-	Active particle fraction of carbonator inventory
G_s	kg/m^2s	Solid circulation mass flow between reactors
H	m	Bed height
h	kJ/kg	Specific enthalpy
HR	-	Heat ratio
k	-	Constant for the decrease of reactivity of limestone
k_s	$1/s$	Reaction rate constant for carbonation
LR	$mole_{Ca}/mole_{CO2}$	Solid looping ratio
M	kg/kmole	Molar mass
$\dot{m}$	kg/h	Mass flow
MUR	$mole_{Ca}/mole_{CO2}$	Make-up ratio
n	mole	Molar mass
N	-	Number of cycles
$O_{2,s}$	$kg_{O2}/kg_{CO2,capt}$	Specific oxygen consumption
P	kW	Power
p	mbar	Pressure
$\dot{Q}$	kW	Heat flow
Q_s	$MJ/kg_{CO2,capt}$	Specific heat demand
T	°C	Temperature
t	s	Time
t^*	s	Characteristic time of the carbonation reaction
u^*	-	Dimensionless velocity
u_0	m/s	Superficial gas velocity
u_t	m/s	Terminal velocity
V	m^3	Volume
$\dot{V}$	m^3/h	Volume flow

W_s	kg/m^2	Specific solid inventory
X	mole$_{CaCO3}$/mole$_{Ca}$	Molar conversion
x	-	Variable
X_{avg}	mole$_{CaCO3}$/mole$_{Ca}$	Average molar conversion
X_{calc}	mole$_{CaCO3}$/mole$_{Ca}$	Molar conversion at the calciner outlet
X_{carb}	mole$_{CaCO3}$/mole$_{Ca}$	Molar conversion at the carbonator outlet
X_{cv}	%	Cone valve position
X_N	mole$_{CaCO3}$/mole$_{Ca}$	Molar conversion after N cylces
X_r	mole$_{CaCO3}$/mole$_{Ca}$	Residual molar activity of the sorbent
y	wt.%	Mass fraction

Greek symbols

$\Delta\eta_0$	%.-points	Efficiency penalty by make-up
$\Delta\eta_{ASU}$	%.-points	Efficiency penalty by ASU
$\Delta\eta_{CaL}$	%.-points	Efficiency penalty of the CaL process
ΔH	kJ/mol	Reaction enthalpie
Δp	mbar	Pressure drop of the fluidized bed
ε	-	Void fraction
μ	kg/ms	Dynamic viscosity
ν	vol.%	Volumetric fraction
ϕ	-	Sphericity
φ	-	Gas–solid contacting effectivity factor
ρ	kg/m^3	Density
σ	-	Standard deviation
τ_{active}	s	Active space time

Constants and dimensionless numbers

g	9.81 m/s^2	Gravity constant
Re	-	Reynolds number

Indices

0	Make-up
ash	Ash
calc	Calciner
capt	Captured
carb	Carbonator
coal	Coal

cool	Cooling
corr	Corrected
db	Dry basis
el	Electrical
eq	Equilibrium
f	Fluid
flush	Flushing
g	Gas Phase
i,j	Counter
in	Inlet
loss	Loss
max	Maximum
mf	Minumum fluidization
out	Outlet
P	Particle
rec	Recirculation
s	Solid phase
th	Thermal
tot	Total

Chemical compounds

Al_2O_3	Aluminium oxide
Ar	Argon
C	Carbon
C_3H_8	Propane
Ca	Calcium
$CaCO_3$	Calcium carbonate
CaO	Calcium oxide
$CaSO_4$	Calcium sulphate
CH_4	Methane
CO	Carbon monoxide
CO_2	Carbon dioxide
Fe_2O_3	Hematite
H_2	Hydrogen
H_2O	Water
$MgCO_3$	Magnesium Carbonate
N_2	Nitrogen
NO	Nitrogen oxide

NO_x	Nitrogen oxides
O_2	Oxygen
S	Sulphur
SiO_2	Silicon dioxide
SO_2	Sulphur dioxide
SO_x	Sulphur oxides

Abbreviations

ASPEN PLUS	Commercial Process Simulation Software
ASU	Air Separation Unit
BET	Brunaur-Emmett-Teller(-Analysis)
BFB	Bubbling Fluidized Bed
BJH	Barrett-Joyner-Halenda(-Analysis)
CaL	Carbonate/Calcium Looping
CANMET	Working group of Canadian Energy Technology Centre (CETC)
CCS	Carbon Capture and Storage
CCU	Carbon Capture and Utilization
CFB	Circulating Fluidized Bed
CFD	Computational Fluid Dynamics
CLC	Chemical Looping Combustion
CSIC-INCAR	Instituto Nacional del Carbón - Consejo Superior de Investigaciones Científicas
CV	Cone Valve
EB	Entrained Bed
EC	El Cerrejon, coal type
EGR	Enhanced Gas Recovery
EOR	Enhanced Oil Recovery
FFB	Fast Fluidized Bed
FGD	Flue Gas Desulphurization
FORTRAN	Formular Transformation, Programming Language
GPU	Gas Processing Unit
HC	Hard Coal
HE	Heat Exchanger
IEA	International Energy Agency
IEAGHG	IEA Greenhouse Gas R&D Programme
IF	Istein Fine Limestone
IGCC	Integrated Gasification Combined Cycle
IHCaL	Indirectly Heated Carbonate Looping
IPCC	Intergovernmental Panel on Climate Change

ITRI	Industrial Technology Research Institute of Taiwan
LEG	Lignite Energy Grained
LEP	Lignite Energy Pulverized
LHV	Lower Heating Value
LS	Loop seal
MB	Moving Bed
MEA	Mono Ethylen Amine
MF	Messinghausen Fine Limestone
OCC	Other Calcium Compounds
OP	Operating Point
PSD	Particle Size Distribution
RDF	Refuse-Derived Fuel
SCARLET	Project: Scale-up of Calcium Carbonate Looping Technology for Efficient CO_2 Capture from Power and Industrial Plants
SCR	Selective Catalytic Reaction
SFGD	Seawater Flue Gas Desulphurization
TFB	Turbulent Fluidized Bed
TGA	Thermogravimetric Analysis
TRL	Technology Readiness Level
UHS	US High Sulphur, coal type
XRF	X-Ray Fluorescence

Bibliography

[1] S. Rackley, *Carbon Capture and Storage*. Elsevier Science, 2017.

[2] L. Bernstein, P. Bosch, O. Canziani, Z. Chen, R. Christ, and K. Riahi, "Climate Change 2007: Synthesis Report - An Assessment of the Intergovernmental Panel on Climate Change (IPCC)," Report, IPCC, 2008.

[3] O. Edenhofer, R. Pichs-Madruga, Y. Sokona, E. Farahani, S. Kadner, K. Seyboth, A. Adler, I. Baum, S. Brunner, P. Eickemeier, B. Kriemann, J. Savolainen, S. Schlömer, C. von Stechow, T. Zwickel, and J. Minx, "Climate Change 2014: Mitigation of Climate Change. Contribution of Working Group III to the Fifth Assessment Report of the Intergovernmental Panel on Climate Change," Report, IPCC, 2014.

[4] International Energy Agengy (IEA), *World Energy Outlook 2015*. 2015.

[5] COP 2015, "FCCC/CP/2015/10/Add.2: Report of the Conference of the Parties on its twenty-first session, held in Paris from 30 November to 13 December 2015, Part two: Action taken by the Conference of the Parties at its twenty-first session," Report, United Nations.

[6] International Energy Agengy (IEA), *World Energy Outlook 2017*. 2017.

[7] J. C. Abanades, B. Arias, A. Lyngfelt, T. Mattisson, D. E. Wiley, H. Li, M. T. Ho, E. Mangano, and S. Brandani, "Emerging CO_2 capture systems," *International Journal of Greenhouse Gas Control*, vol. 40, pp. 126–166, 2015.

[8] IEAGHG, "Assessment of emerging CO_2 capture technologies and their potential to reduce costs, 2014/TR4," Report, 2014.

[9] B. Metz, O. Davidson, H. de Coninck, M. Loos, and L. Meyer, "IPCC Special Report on Carbon Dioxide Capture and Storage," Report 9780521685511, 2005.

[10] W. H. Chen, S. M. Chen, and C. I. Hung, "Carbon dioxide capture by single droplet using Selexol, Rectisol and water as absorbents: A theoretical approach," *Applied Energy*, vol. 111, pp. 731–741, 2013.

[11] H. Weiss, "Rectisol wash for purification of partial oxidation gases," *Gas Separation &Purification*, vol. 2, no. 4, pp. 171–176, 1988.

[12] C. Descamps, C. Bouallou, and M. Kanniche, "Efficiency of an Integrated Gasification Combined Cycle (IGCC) power plant including CO_2 removal," *Energy*, vol. 33, no. 6, pp. 874–881, 2008.

[13] M. Geuss, "$7.5 billion Kemper power plant suspends coal gasification," 2017.

[14] A. B. Rao and E. S. Rubin, "A technical, economic, and environmental assessment of amine-based CO_2 capture technology for power plant greenhouse gas control," *Environmental Science & Technology*, vol. 36, no. 20, pp. 4467–4475, 2002.

[15] B. Dutcher, M. H. Fan, and A. G. Russell, "Amine-Based CO_2 Capture Technology Development from the Beginning of 2013-A Review," *Acs Applied Materials & Interfaces*, vol. 7, no. 4, pp. 2137–2148, 2015.

[16] V. Darde, K. Thomsen, W. J. M. van Well, and E. H. Stenby, "Chilled ammonia process for CO_2 capture," *International Journal of Greenhouse Gas Control*, vol. 4, no. 2, pp. 131–136, 2010.

[17] J. Ströhle, A. Galloy, and B. Epple, "Feasibility study on the carbonate looping process for post-combustion CO_2 capture from coal-fired power plants," *Energy Procedia*, vol. 1, no. 1, pp. 1313–1320, 2009.

[18] IEAGHG, "Integrated Carbon Capture and Storage Project at SaskPowers Boundary Dam Power Station, 2015/06," Report, 2015.

[19] H. C. Mantripragada, H. B. Zhai, and E. S. Rubin, "Boundary Dam or Petra Nova - Which is a better model for CCS energy supply?," *International Journal of Greenhouse Gas Control*, vol. 82, pp. 59–68, 2019.

[20] R. Stanger, T. Wall, R. Sporl, M. Paneru, S. Grathwohl, M. Weidmann, G. Schefflmecht, D. Mc-Donald, K. Myohanen, J. Ritvanen, S. Rahiala, T. Hyppanen, J. Mletzko, A. Kather, and S. Santos, "Oxyfuel combustion for CO_2 capture in power plants," *International Journal of Greenhouse Gas Control*, vol. 40, pp. 55–125, 2015.

[21] B. J. P. Buhre, L. K. Elliott, C. D. Sheng, R. P. Gupta, and T. F. Wall, "Oxy-fuel combustion technology for coal-fired power generation," *Progress in Energy and Combustion Science*, vol. 31, no. 4, pp. 283–307, 2005.

[22] P. N. Dyer, R. E. Richards, S. L. Russek, and D. M. Taylor, "Ion transport membrane technology for oxygen separation and syngas production," *Solid State Ionics*, vol. 134, no. 1-2, pp. 21–33, 2000.

[23] A. Lyngfelt, B. Leckner, and T. Mattisson, "A fluidized-bed combustion process with inherent CO_2 separation; application of chemical-looping combustion," *Chemical Engineering Science*, vol. 56, no. 10, pp. 3101–3113, 2001.

[24] M. Anheden, U. Burchhardt, H. Ecke, R. Faber, O. Jidinger, R. Giering, H. Kass, S. Lysk, E. Ramström, and J. Yan, "Overview of operational experience and results from test activities in Vattenfall's 30 MW_{th} oxyfuel pilot plant in Schwarze Pumpe," *Energy Procedia*, vol. 4, pp. 941–950, 2011.

[25] S. Martens, A. Liebscher, F. Möller, J. Henninges, T. Kempka, S. Lüth, B. Norden, B. Prevedel, A. Szizybalski, M. Zimmer, M. Kühn, and K. Group, "CO$_2$ Storage at the Ketzin Pilot Site, Germany: Fourth Year of Injection, Monitoring, Modelling and Verification," *Energy Procedia*, vol. 37, pp. 6434–6443, 2013.

[26] A. Komaki, T. Gotou, T. Uchida, T. Yamada, T. Kiga, and C. Spero, "Operation Experiences of Oxyfuel Power Plant in Callide Oxyfuel Project," *Energy Procedia*, vol. 63, pp. 490–496, 2014.

[27] T. Uchida, T. Goto, T. Yamada, T. Kiga, and C. Spero, "Oxyfuel Combustion as CO$_2$ Capture Technology Advancing for Practical use - callide Oxyfuel Project," *Energy Procedia*, vol. 37, pp. 1471–1479, 2013.

[28] Network, European CCS Demonstration Project, "Lessons learned from the Jänschwalde project - Summary report," Report, 2012.

[29] J. Hilz, M. Helbig, M. Haaf, A. Daikeler, J. Ströhle, and B. Epple, "Investigation of the fuel influence on the carbonate looping process in 1 MW$_{th}$ scale," *Fuel Processing Technology*, vol. 169, pp. 170–177, 2018.

[30] B. Arias, M. E. Diego, J. C. Abanades, M. Lorenzo, L. Diaz, D. Martinez, J. Alvarez, and A. Sanchez-Biezma, "Demonstration of steady state CO$_2$ capture in a 1.7 MW$_{th}$ calcium looping pilot," *International Journal of Greenhouse Gas Control*, vol. 18, pp. 237–245, 2013.

[31] P. Ohlemüller, J. P. Busch, M. Reitz, J. Ströhle, and B. Epple, "Chemical-Looping Combustion of Hard Coal: Autothermal Operation of a 1 MW$_{th}$ Pilot Plant," *Journal of Energy Resources Technology-Transactions of the ASME*, vol. 138, no. 4, 2016.

[32] P. Ohlemüller, J. Ströhle, and B. Epple, "Chemical looping combustion of hard coal and torrefied biomass in a 1 MW$_{th}$ pilot plant," *International Journal of Greenhouse Gas Control*, vol. 65, pp. 149–159, 2017.

[33] P. Ohlemüller, M. Reitz, J. Ströhle, and B. Epple, "Investigation of chemical looping combustion of natural gas at 1 MW$_{th}$ scale," *Proceedings of the Combustion Institute*, vol. 37, no. 4, pp. 4353–4360, 2019.

[34] G. Pipitone and O. Bolland, "Power generation with CO$_2$ capture: Technology for CO$_2$ purification," *International Journal of Greenhouse Gas Control*, vol. 3, no. 5, pp. 528–534, 2009.

[35] P. Noothout, F. Wiersma, O. Hurtado, D. Macdonald, J. Kemper, and K. van Alphen, "CO$_2$ Pipeline infrastructure - lessons learnt," *Energy Procedia*.

[36] T. Suzuki, M. Toriumi, T. Sakemi, N. Masui, S. Yano, H. Fujita, and H. Furukawa, "Conceptual Design of CO$_2$ Transportation System for CCS," *Energy Procedia*, vol. 37, pp. 2989–2996, 2013.

[37] T. Vangkilde-Pedersen, K. L. Anthonsen, N. Smith, K. Kirk, F. neele, B. van der Meer, Y. Le Gallo, D. Bossie-Codreanu, A. Wojcicki, Y.-M. Le Nindre, C. Hendriks, F. Dalhoff, and N. Peter Christensen, "Assessing European capacity for geological storage of carbon dioxide–the EU GeoCapacity project," *Energy Procedia*, vol. 1, no. 1, pp. 2663–2670, 2009.

[38] H. Rütters, S. Stadler, R. Bassler, D. Bettge, S. Jeschke, A. Kather, C. Lempp, U. Lubenau, C. Ostertag-Henning, S. Schmitz, S. Schutz, S. Waldmann, and C. Team, "Towards an optimization of the CO_2 stream composition-A whole-chain approach," *International Journal of Greenhouse Gas Control*, vol. 54, pp. 682–701, 2016.

[39] L. S. Melzer, "Carbon Dioxide Enhanced Oil Recovery (CO_2 EOR): Factors Involved in Adding Carbon Capture, Utilization and Storage (CCUS) to Enhanced Oil Recovery," Report, Melzer Consulting - CO_2 Consultant, 2012.

[40] C. H. Huang and C. S. Tan, "A Review: CO_2 Utilization," *Aerosol and Air Quality Research*, vol. 14, no. 2, pp. 480–499, 2014.

[41] Association, European Industrial Gases (EIGA), "Carbon Dioxide Source Qualification, Quality Standards and Verification (IGC Doc 70/08/E)," Report, EIGA, 2008.

[42] M. Chen, Z. L. Guo, J. Zheng, F. L. Jing, and W. Chu, "CO_2 selective hydrogenation to synthetic natural gas (SNG) over four nano-sized Ni/ZrO_2 samples: ZrO_2 crystalline phase & treatment impact," *Journal of Energy Chemistry*, vol. 25, no. 6, pp. 1070–1077, 2016.

[43] L. Zhou, "Progress and problems in hydrogen storage methods," *Renewable & Sustainable Energy Reviews*, vol. 9, no. 4, pp. 395–408, 2005.

[44] E. S. Rubin, G. Booras, J. Davison, C. Ekstrom, M. Matuszewski, S. McCoy, and C. Short, "Toward a common method of cost estimation for CO_2 capture and storage at fossil fuel power plants," Report, Task Force on CCS Costing Methods, 2013.

[45] M. Junk, M. Reitz, J. Ströhle, and B. Epple, "Technical and Economical Assessment of the Indirectly Heated Carbonate Looping Process," *Journal of Energy Resources Technology-Transactions of the ASME*, vol. 138, no. 4, 2016.

[46] IEAGHG, "CO_2 Capture at Coal Based Power and Hydrogen Plants, 2014/03," Report, 2014.

[47] A. Rolfe, Y. Huang, M. Haaf, S. Rezvani, D. McIlveen-Wright, and N. J. Hewitt, "Integration of the calcium carbonate looping process into an existing pulverized coal-fired power plant for CO_2 capture: Techno-economic and environmental evaluation," *Applied Energy*, vol. 222, pp. 169–179, 2018.

[48] M. Finkenrath, "Cost and Performance of Carbon Dioxide Capture from Power Generation," Report, International Energy Agency, 2011.

[49] J. C. Abanades, E. J. Anthony, D. Y. Lu, C. Salvador, and D. Alvarez, "Capture of CO_2 from combustion gases in a fluidized bed of CaO," *Aiche Journal*, vol. 50, no. 7, pp. 1614–1622, 2004.

[50] J. C. Abanades and D. Alvarez, "Conversion limits in the reaction of CO_2 with lime," *Energy & Fuels*, vol. 17, no. 2, pp. 308–315, 2003.

[51] T. Shimizu, T. Hirama, H. Hosoda, K. Kitano, M. Inagaki, and K. Tejima, "A twin fluid-bed reactor for removal of CO_2 from combustion processes," *Chemical Engineering Research & Design*, vol. 77, no. A1, pp. 62–68, 1999.

[52] J. C. Abanades, E. J. Anthony, J. S. Wang, and J. E. Oakey, "Fluidized bed combustion systems integrating CO_2 capture with CaO," *Environmental Science & Technology*, vol. 39, no. 8, pp. 2861–2866, 2005.

[53] J. R. Fernandez and J. C. Abanades, "CO_2 capture from the calcination of $CaCO_3$ using iron oxide as heat carrier," *Journal of Cleaner Production*, vol. 112, pp. 1211–1217, 2016.

[54] I. Martinez, R. Murillo, G. Grasa, N. Rodriguez, and J. C. Abanades, "Conceptual design of a three fluidised beds combustion system capturing CO_2 with CaO," *International Journal of Greenhouse Gas Control*, vol. 5, no. 3, pp. 498–504, 2011.

[55] I. Martinez, G. Grasa, J. Parkkinen, T. Tynjala, T. Hyppanen, R. Murillo, and M. C. Romano, "Review and research needs of Ca-Looping systems modelling for post-combustion CO_2 capture applications," *International Journal of Greenhouse Gas Control*, vol. 50, pp. 271–304, 2016.

[56] M. Junk, M. Reitz, J. Ströhle, and B. Epple, "Thermodynamic Evaluation and Cold Flow Model Testing of an Indirectly Heated Carbonate Looping Process," *Chemical Engineering & Technology*, vol. 36, no. 9, pp. 1479–1487, 2013.

[57] G. D. Silcox, J. C. Kramlich, and D. W. Pershing, "A Mathematical-Model for the Flash Calcination of Dispersed $CaCO_3$ and $Ca(OH)_2$ Particles," *Industrial & Engineering Chemistry Research*, vol. 28, no. 2, pp. 155–160, 1989.

[58] F. Garcia-Labiano, A. Abad, L. F. de Diego, P. Gayan, and J. Adanez, "Calcination of calcium-based sorbents at pressure in a broad range of CO_2 concentrations," *Chemical Engineering Science*, vol. 57, no. 13, pp. 2381–2393, 2002.

[59] D. Y. Lu, R. W. Hughes, and E. J. Anthony, "Ca-based sorbent looping combustion for CO_2 capture in pilot-scale dual fluidized beds," *Fuel Processing Technology*, vol. 89, no. 12, pp. 1386–1395, 2008.

[60] N. Y. Hu and A. W. Scaroni, "Calcination of pulverized limestone particles under furnace injection conditions," *Fuel*, vol. 75, no. 2, pp. 177–186, 1996.

[61] P. Sun, J. R. Grace, C. J. Lim, and E. J. Anthony, "Determination of intrinsic rate constants of the CaO-CO_2 reaction," *Chemical Engineering Science*, vol. 63, no. 1, pp. 47–56, 2008.

[62] G. S. Grasa and J. C. Abanades, "CO$_2$ capture capacity of CaO in long series of carbonation/calcination cycles," *Industrial & Engineering Chemistry Research*, vol. 45, no. 26, pp. 8846–8851, 2006.

[63] P. S. Fennell, R. Pacciani, J. S. Dennis, J. F. Davidson, and A. N. Hayhurst, "The effects of repeated cycles of calcination and carbonation on a variety of different limestones, as measured in a hot fluidized bed of sand," *Energy & Fuels*, vol. 21, no. 4, pp. 2072–2081, 2007.

[64] S. K. Bhatia and D. D. Perlmutter, "Effect of the Product Layer on the Kinetics of the CO$_2$-Lime Reaction," *Aiche Journal*, vol. 29, no. 1, pp. 79–86, 1983.

[65] D. Alvarez and J. C. Abanades, "Pore-size and shape effects on the recarbonation performance of calcium oxide submitted to repeated calcination/recarbonation cycles," *Energy & Fuels*, vol. 19, no. 1, pp. 270–278, 2005.

[66] G. S. Grasa, J. C. Abanades, M. Alonso, and B. Gonzalez, "Reactivity of highly cycled particles of CaO in a carbonation/calcination loop," *Chemical Engineering Journal*, vol. 137, no. 3, pp. 561–567, 2008.

[67] F. Donat, N. H. Florin, E. J. Anthony, and P. S. Fennell, "Influence of High-Temperature Steam on the Reactivity of CaO Sorbent for CO$_2$ Capture," *Environmental Science & Technology*, vol. 46, no. 2, pp. 1262–1269, 2012.

[68] D. Y. Lu, R. W. Hughes, E. J. Anthony, and V. Manovic, "Sintering and Reactivity of CaCO$_3$-Based Sorbents for In Situ CO$_2$ Capture in Fluidized Beds under Realistic Calcination Conditions," *Journal of Environmental Engineering-ASCE*, vol. 135, no. 6, pp. 404–410, 2009.

[69] R. H. Borgwardt, "Sintering of nascent calcium oxide," *Chemical Engineering Science*, vol. 44, no. 1, pp. 53–60, 1989.

[70] V. Manovic and E. J. Anthony, "Carbonation of CaO-Based Sorbents Enhanced by Steam Addition," *Industrial & Engineering Chemistry Research*, vol. 49, no. 19, pp. 9105–9110, 2010.

[71] B. Arias, G. Grasa, J. C. Abanades, V. Manovic, and E. J. Anthony, "The Effect of Steam on the Fast Carbonation Reaction Rates of CaO," *Industrial & Engineering Chemistry Research*, vol. 51, no. 5, pp. 2478–2482, 2012.

[72] R. T. Symonds, D. Y. Lu, V. Manovic, and E. J. Anthony, "Pilot-Scale Study of CO$_2$ Capture by CaO-Based Sorbents in the Presence of Steam and SO$_2$," *Industrial & Engineering Chemistry Research*, vol. 51, no. 21, pp. 7177–7184, 2012.

[73] R. T. Symonds, D. Y. Lu, R. W. Hughes, E. J. Anthony, and A. Macchi, "CO$_2$ Capture from Simulated Syngas via Cyclic Carbonation/Calcination for a Naturally Occurring Limestone: Pilot-Plant Testing," *Industrial & Engineering Chemistry Research*, vol. 48, no. 18, pp. 8431–8440, 2009.

[74] J. Blamey, E. J. Anthony, J. Wang, and P. S. Fennell, "The calcium looping cycle for large-scale CO_2 capture," *Progress in Energy and Combustion Science*, vol. 36, no. 2, pp. 260–279, 2010.

[75] B. R. Stanmore and P. Gilot, "Review-calcination and carbonation of limestone during thermal cycling for CO_2 sequestration," *Fuel Processing Technology*, vol. 86, no. 16, pp. 1707–1743, 2005.

[76] S. Champagne, D. Y. Lu, R. T. Symonds, A. Macchi, and E. J. Anthony, "The effect of steam addition to the calciner in a calcium looping pilot plant," *Powder Technology*, vol. 290, pp. 114–123, 2016.

[77] S. Champagne, D. Y. Lu, A. Macchi, R. T. Symonds, and E. J. Anthony, "Influence of Steam Injection during Calcination on the Reactivity of CaO-Based Sorbent for Carbon Capture," *Industrial & Engineering Chemistry Research*, vol. 52, no. 6, pp. 2241–2246, 2013.

[78] M. E. Diego, B. Arias, M. Alonso, and J. C. Abanades, "The impact of calcium sulfate and inert solids accumulation in post-combustion calcium looping systems," *Fuel*, vol. 109, pp. 184–190, 2013.

[79] L. M. Romeo, Y. Lara, P. Lisbona, and J. M. Escosa, "Optimizing make-up flow in a CO_2 capture system using CaO," *Chemical Engineering Journal*, vol. 147, no. 2-3, pp. 252–258, 2009.

[80] H. J. Ryu, J. R. Grace, and C. J. Lim, "Simultaneous CO_2/SO_2 capture characteristics of three limestones in a fluidized-bed reactor," *Energy & Fuels*, vol. 20, no. 4, pp. 1621–1628, 2006.

[81] A. Coppola, F. Montagnaro, P. Salatino, and F. Scala, "Fluidized bed calcium looping: The effect of SO_2 on sorbent attrition and CO_2 capture capacity," *Chemical Engineering Journal*, vol. 207, pp. 445–449, 2012.

[82] V. Manovic and E. J. Anthony, "Steam reactivation of spent CaO-based sorbent for multiple CO_2 capture cycles," *Environmental Science & Technology*, vol. 41, no. 4, pp. 1420–1425, 2007.

[83] M. V. Iyer, H. Gupta, B. B. Sakadjian, and L. S. Fan, "Multicyclic study on the simultaneous carbonation and sulfation of high-reactivity CaO," *Industrial & Engineering Chemistry Research*, vol. 43, no. 14, pp. 3939–3947, 2004.

[84] G. S. Grasa, M. Alonso, and J. C. Abanades, "Sulfation of CaO particles in a carbonation/calcination loop to capture CO_2," *Industrial & Engineering Chemistry Research*, vol. 47, no. 5, pp. 1630–1635, 2008.

[85] J. M. Huang and K. E. Daugherty, "Lithium-Carbonate Enhancement of the Calcination of Calcium-Carbonate - Proposed Extended-Shell Model," *Thermochimica Acta*, vol. 118, pp. 135–141, 1987.

[86] J. M. Huang and K. E. Daugherty, "Inhibition of the Calcination of Calcium-Carbonate," *Thermochimica Acta*, vol. 130, pp. 173–176, 1988.

[87] E. J. Anthony, A. P. Iribarne, J. V. Iribarne, and L. Jia, "Reuse of landfilled FBC residues," *Fuel*, vol. 76, no. 7, pp. 603–606, 1997.

[88] Y. C. Ray, T. S. Jiang, and T. L. Jiang, "Particle-Population Model for a Fluidized-Bed with Attrition," *Powder Technology*, vol. 52, no. 1, pp. 35–48, 1987.

[89] F. Scala, A. Cammarota, R. Chirone, and P. Salatino, "Comminution of limestone during batch fluidized-bed calcination and sulfation," *Aiche Journal*, vol. 43, no. 2, pp. 363–373, 1997.

[90] F. Scala, P. Salatino, R. Boerefijn, and M. Ghadiri, "Attrition of sorbents during fluidized bed calcination and sulphation," *Powder Technology*, vol. 107, no. 1-2, pp. 153–167, 2000.

[91] F. Scala, F. Montagnaro, and P. Salatino, "Attrition of limestone by impact loading in fluidized beds," *Energy & Fuels*, vol. 21, no. 5, pp. 2566–2572, 2007.

[92] L. Jia, R. Hughes, D. Lu, E. J. Anthony, and I. Lau, "Attrition of calcining limestones in circulating fluidized-bed systems," *Industrial & Engineering Chemistry Research*, vol. 46, no. 15, pp. 5199–5209, 2007.

[93] J. Saastamoinen, T. Pikkarainen, A. Tourunen, M. Rasanen, and T. Jantti, "Model of fragmentation of limestone particles during thermal shock and calcination in fluidised beds," *Powder Technology*, vol. 187, no. 3, pp. 244–251, 2008.

[94] F. Montagnaro, P. Salatino, and F. Scala, "The influence of temperature on limestone sulfation and attrition under fluidized bed combustion conditions," *Experimental Thermal and Fluid Science*, vol. 34, no. 3, pp. 352–358, 2010.

[95] M. Kraume, *Transportvorgänge in der Verfahrenstechnik: Grundlagen und apparative Umsetzungen*. Springer Berlin Heidelberg, 2013.

[96] O. Levenspiel, *Chemical Reaction Engineering*. John Wiley and Sons, 3rd ed., 1999.

[97] D. Kunii and O. Levenspiel, *Fluidization Engineering*. Boston: Butterworth-Heinemann, 2nd ed., 1991.

[98] J. R. Grace, "Contacting modes and behaviour classification of gas—solid and other two-phase suspensions," *The Canadian Journal of Chemical Engineering*, vol. 64, no. 3, pp. 353–363, 1986.

[99] D. Geldart, "Types of Gas Fluidization," *Powder Technology*, vol. 7, no. 5, pp. 285–292, 1973.

[100] D. Geldart, *Gas Fluidization Technology*. John Wiley and Sons.

[101] H. A. Jakobsen, *Chemical Reactor Modeling: Multiphase Reactive Flows*. Berlin, Heidelberg: Springer Publishing Company, Incorporated, 2nd ed., 2014.

[102] H. Dieter, M. Beirow, D. Schweitzer, C. Hawthorne, and G. Scheffknecht, "Efficiency and flexibility potential of Calcium Looping CO_2 Capture," *Energy Procedia*, vol. 63, pp. 2129–2137, 2014.

[103] N. Rodriguez, M. Alonso, G. Grasa, and J. C. Abanades, "Heat requirements in a calciner of $CaCO_3$ integrated in a CO_2 capture system using CaO," *Chemical Engineering Journal*, vol. 138, no. 1-3, pp. 148–154, 2008.

[104] N. Rodríguez, M. Alonso, J. C. Abanades, A. Charitos, C. Hawthorne, G. Scheffknecht, D. Y. Lu, and E. J. Anthony, "Comparison of experimental results from three dual fluidized bed test facilities capturing CO_2 with CaO," *Energy Procedia*, vol. 4, pp. 393–401, 2011.

[105] A. Charitos, N. Rodriguez, C. Hawthorne, M. Alonso, M. Zieba, B. Arias, G. Kopanakis, G. Scheffknecht, and J. C. Abanades, "Experimental Validation of the Calcium Looping CO_2 Capture Process with Two Circulating Fluidized Bed Carbonator Reactors," *Industrial & Engineering Chemistry Research*, vol. 50, no. 16, pp. 9685–9695, 2011.

[106] N. Rodriguez, M. Alonso, and J. C. Abanades, "Average activity of CaO particles in a calcium looping system," *Chemical Engineering Journal*, vol. 156, no. 2, pp. 388–394, 2010.

[107] F. Fang, Z.-S. Li, and N.-S. Cai, "Continuous CO_2 Capture from Flue Gases Using a Dual Fluidized Bed Reactor with Calcium-Based Sorbent," *Industrial & Engineering Chemistry Research*, vol. 48, no. 24, pp. 11140–11147, 2009.

[108] A. Charitos, C. Hawthorne, A. R. Bidwe, S. Sivalingam, A. Schuster, H. Spliethoff, and G. Scheffknecht, "Parametric investigation of the calcium looping process for CO_2 capture in a 10 kW$_{th}$ dual fluidized bed," *International Journal of Greenhouse Gas Control*, vol. 4, no. 5, pp. 776–784, 2010.

[109] G. Duelli, A. Charitos, M. E. Diego, E. Stavroulakis, H. Dieter, and G. Scheffknecht, "Investigations at a 10 kW$_{th}$ calcium looping dual fluidized bed facility: Limestone calcination and CO_2 capture under high CO_2 and water vapor atmosphere," *International Journal of Greenhouse Gas Control*, vol. 33, pp. 103–112, 2015.

[110] A. Cotton, K. N. Finney, K. Patchigolla, R. E. A. Eatwell-Hall, J. E. Oakey, J. Swithenbank, and V. Sharifi, "Quantification of trace element emissions from low-carbon emission energy sources: (I) Ca-looping cycle for post-combustion CO_2 capture and (II) fixed bed, air blown down-draft gasifier," *Chemical Engineering Science*, vol. 107, pp. 13–29, 2014.

[111] J. C. Abanades, M. Alonso, N. Rodríguez, B. González, G. Grasa, and R. Murillo, "Capturing CO_2 from combustion flue gases with a carbonation calcination loop. Experimental results and process development," *Energy Procedia*, vol. 1, no. 1, pp. 1147–1154, 2009.

[112] B. González, J. C. Abanades, and M. Alonso, "Sorbent attrition in a carbonation/calcination pilot plant for capturing CO_2 from flue gases," *Fuel*, vol. 89, no. 10, pp. 2918–2924, 2010.

[113] N. Rodriguez, M. Alonso, and J. C. Abanades, "Experimental Investigation of a Circulating Fluidized-Bed Reactor to Capture CO_2 with CaO," *Aiche Journal*, vol. 57, no. 5, pp. 1356–1366, 2011.

[114] R. W. Hughes, D. Y. Lu, E. J. Anthony, and A. Macchi, "Design, process simulation and construction of an atmospheric dual fluidized bed combustion system for in situ CO_2 capture using high-temperature sorbents," *Fuel Processing Technology*, vol. 86, no. 14-15, pp. 1523–1531, 2005.

[115] W. Wang, S. Ramkumar, S. G. Li, D. Wong, M. Iyer, B. B. Sakadjian, R. M. Statnick, and L. S. Fan, "Subpilot Demonstration of the Carbonation-Calcination Reaction (CCR) Process: High-Temperature CO_2 and Sulfur Capture from Coal-Fired Power Plants," *Industrial & Engineering Chemistry Research*, vol. 49, no. 11, pp. 5094–5101, 2010.

[116] C. Hawthorne, H. Dieter, A. Bidwe, A. Schuster, G. Scheffknecht, S. Unterberger, and M. Käß, "CO_2 capture with CaO in a 200 kW_{th} dual fluidized bed pilot plant," *Energy Procedia*, vol. 4, pp. 441–448, 2011.

[117] H. Dieter, C. Hawthorne, M. Zieba, and G. Scheffknecht, "Progress in Calcium Looping Post Combustion CO_2 Capture: Successful Pilot Scale Demonstration," *Energy Procedia*, vol. 37, pp. 48–56, 2013.

[118] M. Hornberger, R. Spörl, and G. Scheffknecht, "Calcium Looping for CO_2 Capture in Cement Plants – Pilot Scale Test," *Energy Procedia*, vol. 114, pp. 6171–6174, 2017.

[119] M. Alonso, M. E. Diego, C. Perez, J. R. Chamberlain, and J. C. Abanades, "Biomass combustion with in situ CO_2 capture by CaO in a 300 kW_{th} circulating fluidized bed facility," *International Journal of Greenhouse Gas Control*, vol. 29, pp. 142–152, 2014.

[120] M. E. Diego and M. Alonso, "Operational feasibility of biomass combustion with in situ CO_2 capture by CaO during 360 h in a 300 kW_{th} calcium looping facility," *Fuel*, vol. 181, pp. 325–329, 2016.

[121] M. Reitz, M. Junk, J. Ströhle, and B. Epple, "Design and erection of a 300 kW_{th} indirectly heated carbonate looping test facility," *Energy Procedia*, vol. 63, pp. 2170–2177, 2014.

[122] M. Reitz, M. Junk, J. Ströhle, and B. Epple, "Design and operation of a 300 $kW_{(th)}$ indirectly heated carbonate looping pilot plant," *International Journal of Greenhouse Gas Control*, vol. 54, pp. 272–281, 2016.

[123] J. Kremer, A. Galloy, J. Ströhle, and B. Epple, "Continuous CO_2 Capture in a 1-MW_{th} Carbonate Looping Pilot Plant," *Chemical Engineering & Technology*, vol. 36, no. 9, pp. 1518–1524, 2013.

[124] J. Ströhle, M. Junk, J. Kremer, A. Galloy, and B. Epple, "Carbonate looping experiments in a 1 MW_{th} pilot plant and model validation," *Fuel*, vol. 127, pp. 13–22, 2014.

[125] M. Helbig, J. Hilz, M. Haaf, A. Daikeler, J. Ströhle, and B. Epple, "Long-term carbonate looping testing in a 1 MW_{th} pilot plant with hard coal and lignite," *Energy Procedia*, vol. 114, pp. 179–190, 2017.

[126] J. Hilz, M. Helbig, M. Haaf, A. Daikeler, J. Ströhle, and B. Epple, "Long-term pilot testing of the carbonate looping process in 1 MW_{th} scale," *Fuel*, vol. 210, pp. 892–899, 2017.

[127] M. E. Diego, B. Arias, G. Grasa, J. C. Abanades, L. Diaz, M. Lorenzo, and A. Sanchez-Biezma, "Calcium Looping with enhanced sorbent performance: experimental testing in a large pilot plant," *Energy Procedia*, vol. 63, pp. 2060–2069, 2014.

[128] B. Arias, M. E. Diego, A. Méndez, M. Alonso, and J. C. Abanades, "Calcium looping performance under extreme oxy-fuel combustion conditions in the calciner," *Fuel*, vol. 222, pp. 711–717, 2018.

[129] M. H. Chang, C. M. Huang, W. H. Liu, W. C. Chen, J. Y. Cheng, W. Chen, T. W. Wen, S. Ouyang, C. H. Shen, and H. W. Hsu, "Design and Experimental Investigation of the Calcium Looping Process for 3-kW_{th} and 1.9-MW_{th} Facilities," *Chemical Engineering & Technology*, vol. 36, no. 9, pp. 1525–1532, 2013.

[130] M.-H. Chang, W.-C. Chen, C.-M. Huang, W.-H. Liu, Y.-C. Chou, W.-C. Chang, W. Chen, J.-Y. Cheng, K.-E. Huang, and H.-W. Hsu, "Design and Experimental Testing of a 1.9 MW_{th} Calcium Looping Pilot Plant," *Energy Procedia*, vol. 63, pp. 2100–2108, 2014.

[131] A. Galloy, J. Ströhle, and B. Epple, "Post-combustion CO_2 capture experiments in a 1 MW_{th} carbonate looping pilot," *VGB PowerTech*, no. 6, 2012.

[132] M. Helbig, *Experimentelle Untersuchung des Langzeitverhaltens des Carbonate-Looping-Verfahrens im 1 Megawatt-Technikum*. PhD Thesis, 2019.

[133] Lhoist Recherche et Développement S.A., "D2.1.1 - Results of Attrition Tests," Report, Carbon capture by means of an indirectly heated carbonate looping process (TGC3.01/10), 2011.

[134] Lhoist Western Europe Rheinkalk GmbH Messinghausen, "Product Data Sheet Limestone Grit 0.1 - 0.3 mm, M," Datasheet, 2014.

[135] Lhoist Western Europe Rheinkalk GmbH Istein, "Product Data Sheet Fine Limestone Meal 0.1 - 0.2 mm," Datasheet, 2015.

[136] J. Hilz, M. Haaf, M. Helbig, N. Lindqvist, J. Ströhle, and B. Epple, "Scale-up of the carbonate looping process to a 20 MW_{th} pilot plant based on long-term pilot tests," *International Journal of Greenhouse Gas Control*, vol. 88, pp. 332–341, 2019.

[137] P. Stephan, K. Schaber, K. Stephan, and F. Mayinger, *Thermodynamik: Grundlagen und technische Anwendungen Band 1: Einstoffsysteme*. Springer Berlin Heidelberg, 2013.

[138] M. Reitz, *Experimentelle Untersuchung und Bewertung eines indirekt beheizten Carbonate-Looping-Prozesses*. PhD Thesis, 2017.

[139] G. Duelli, A. Charitos, N. Armbrust, H. Dieter, and G. Scheffknecht, "Analysis of the calcium looping system behavior by implementing simple reactor and attrition models at a 10 kW_{th} dual fluidized bed facility under continuous operation," *Fuel*, vol. 169, pp. 79–86, 2016.

[140] G. Duelli, A. R. Bidwe, I. Papandreou, H. Dieter, and G. Scheffknecht, "Characterization of the oxy-fired regenerator at a 10 kW_{th} dual fluidized bed calcium looping facility," *Applied Thermal Engineering*, vol. 74, pp. 54–60, 2015.

[141] A. Charitos, N. Rodriguez, A. R. Hawthorne, M. Alonso, M. Zieba, and B. Arias, "Validation of a Carbonator Model and Proposal of a Parameter Interdependence Scheme for the Ca-looping CO_2 Capture Process," *Proc. of 21st Int. Conf. on Fluidized Bed Tech., Naples, Italy*, pp. 311–318, 2012.

[142] I. Martinez, G. Grasa, R. Murillo, B. Arias, and J. C. Abanades, "Kinetics of Calcination of Partially Carbonated Particles in a Ca-Looping System for CO_2 Capture," *Energy & Fuels*, vol. 26, no. 2, pp. 1432–1440, 2012.

[143] A. Coppola, F. Scala, G. Itskos, P. Grammelis, H. Pawlak-Kruczek, S. K. Antiohos, P. Salatino, and F. Montagnaro, "Performance of Natural Sorbents during Calcium Looping Cycles: A Comparison between Fluidized Bed and Thermo-Gravimetric Tests," *Energy & Fuels*, vol. 27, no. 10, pp. 6048–6054, 2013.

[144] T. Shimizu, A. Yoshizawa, H. Kim, and L. Y. Li, "Formation of CO and CO_2 in Carbonator and NOx in Regenerator under Calcium Looping Process Conditions," *Journal of Chemical Engineering of Japan*, vol. 49, no. 3, pp. 280–286, 2016.

[145] B. Xue, Y. Yu, J. Chen, X. Luo, and M. Wang, "A comparative study of MEA and DEA for post-combustion CO_2 capture with different process configurations," *International Journal of Coal Science & Technology*, vol. 4, no. 1, pp. 15–24, 2017.

[146] T. Banaszkiewicz, M. Chorowski, and W. Gizicki, "Comparative Analysis of Oxygen Production for Oxy-combustion Application," *Energy Procedia*, vol. 51, pp. 127–134, 2014.

[147] R. W. Hughes, D. Lu, E. J. Anthony, and Y. H. Wu, "Improved long-term conversion of limestone-derived sorbents for in situ capture of CO_2 in a fluidized bed combustor," *Industrial & Engineering Chemistry Research*, vol. 43, no. 18, pp. 5529–5539, 2004.

[148] M. E. Diego, B. Arias, A. Mendez, M. Lorenzo, L. Diaz, A. Sanchez-Biezma, and J. C. Abanades, "Experimental testing of a sorbent reactivation process in la pereda 1.7 mwth calcium looping pilot plant," *International Journal of Greenhouse Gas Control*, vol. 50, pp. 14–22, 2016.

[149] M. Haaf, J. Hilz, M. Helbig, C. Weingärtner, O. Stallmann, J. Ströhle, and B. Epple, "Assessment of the operability of a 20 MW_{th} calcium looping demonstration plant by advanced process modelling," *International Journal of Greenhouse Gas Control*, vol. 75, pp. 224–234, 2018.

[150] Uniper Technologies Limited and Uniper France Power S.A.S., "Deliverable 4.1 - Boundary Conditions of Host Hard Coal Plant," Report, 2016.

[151] M. Haaf, A. Stroh, J. Hilz, M. Helbig, J. Ströhle, and B. Epple, "Process modelling of the calcium looping process and validation against 1 MW_{th} pilot testing," *Energy Procedia*, vol. 114, pp. 167–178, 2017.

[152] A. Stroh, F. Alobaid, M. von Bohnstein, J. Ströhle, and B. Epple, "Numerical CFD simulation of 1 MW_{th} circulating fluidized bed using the coarse grain discrete element method with homogenous drag models and particle size distribution," *Fuel Processing Technology*, vol. 169, pp. 84–93, 2018.

[153] M. Zeneli, A. Nikolopoulos, N. Nikolopoulos, P. Grammelis, and E. Kakaras, "Application of an advanced coupled EMMS-TFM model to a pilot scale CFB carbonator," *Chemical Engineering Science*, vol. 138, pp. 482–498, 2015.

[154] A. Nikolopoulos, A. Stroh, M. Zeneli, F. Alobaid, N. Nikolopoulos, J. Ströhle, S. Karellas, B. Epple, and P. Grammelis, "Numerical investigation and comparison of coarse grain CFD-DEM and TFM in the case of a 1 MW_{th} fluidized bed carbonator simulation," *Chemical Engineering Science*, vol. 163, pp. 189–205, 2017.

[155] M. Zeneli, A. Nikolopoulos, N. Nikolopoulos, P. Grammelis, S. Karellas, and E. Kakaras, "Simulation of the reacting flow within a pilot scale calciner by means of a three phase TFM model," *Fuel Processing Technology*, vol. 162, pp. 105–125, 2017.

List of Figures

List of Tables

Appendix

A.1 Results from Solid Analysis

Tables A.1 to A.6 show the analytical values of sorbent analysis from over 1,000 samples of all test campaigns. The samples were taken at regular intervaLS from the process and hermetically sealed in cans. Subsequently, they were sent them to the laboratories of Lhoist Business Innovation Center, Nivelles in Belgium (test campaigns 1 to 4) and Rheinkalk GmbH, Wülfrath, Germany (test campaigns 5 and 6). The elementary composition was determined by X-ray fluorescence analysis. Furthermore, the surface of the particles was screened by the BET process, the volatile substances by the loss on ignition and the CO_2 content by infrared measurements. The samples are numbered consecutively. The sampling spots from the process are carbonator loop seal (LS 1), the calciner loop seal (LS 2), the heat exchangers (HE_{carb}, HE_{calc}) and the filters ($Filter_{carb}$, $Filter_{calc}$) of both reactors. In addition, a few samples were taken directly from the reactor bottom of the carbonator.

Table A.1: Sample analysis for test campaign 1.

	Date	Time	Spot	CaO wt.%	$CaCO_3$ wt.%	$CaSO_4$ wt.%	Ash wt.%	C wt.%	BET m^2/g	PV cm^3/g
1.66	28.09.15	20:00	HE_{calc}	63.32	25.59	2.14	7.36	3.71	3.33	0.0071
1.75	29.09.15	02:00	HE_{calc}	31.30	2.02	11.60	5.68	5.65	5.41	0.0074
1.91	29.09.15	13:23	LS 1	80.95	5.39	3.18	10.48			
1.92	29.09.15	13:20	LS 2	84.05	2.40	3.21	10.34			
1.95	29.09.15	15:54	LS 1	74.71	11.03	3.56	10.70			
1.96	29.09.15	15:58	LS 2	83.10	2.71	3.52	10.67			
1.99	29.09.15	19:00	LS 1	80.15	4.87	3.47	11.50			
1.100	29.09.15	19:00	LS 2	83.94	2.19	3.23	10.63			
1.103	30.09.15	03:45	LS 2	74.59	8.58	7.60	9.23			
1.104	30.09.15	03:45	LS 1	74.70	8.15	8.95	8.20			
1.105	30.09.15	07:05	LS 2	76.04	3.24	10.13	10.59			
1.106	30.09.15	07:05	LS 1	72.90	8.26	9.81	9.03			
1.109	30.09.15	10:00	LS 2	74.63	3.66	10.57	11.15			
1.110	30.09.15	10:10	LS 1	69.38	7.95	11.44	11.22			
1.115	30.09.15	13:10	LS 1	69.13	6.90	11.77	12.20			
1.116	30.09.15	13:10	LS 2	73.76	2.09	12.16	11.99	0.24	1.32	0.0035
1.121	01.10.15	04:45	HE_{calc}	67.99	15.69	5.98	14.37	2.66	3.58	0.0087
1.126	01.10.15	17:10	LS 1	69.75	14.36	3.59	12.30			
1.127	01.10.15	17:10	LS 2	72.65	7.96	4.01	15.37			

Table A.1: Sample analysis for test campaign 1.

	Date	Time	Spot	CaO wt.%	CaCO$_3$ wt.%	CaSO$_4$ wt.%	Ash wt.%	C wt.%	BET m^2/g	PV cm^3/g
1.129	01.10.15	21:00	LS 1	71.37	13.53	3.34	11.77			
1.130	01.10.15	21:00	LS 2	80.53	4.18	3.35	11.94			
1.157	04.10.15	00:12	LS 1	68.36	10.90	4.79	15.96			
1.158	04.10.15	00:12	LS 2	75.33	4.49	4.43	15.76			
1.160	04.10.15	02:40	Filter$_{carb}$					22.26		
1.161	04.10.15	02:40	Filter$_{calc}$					13.84		
1.162	04.10.15	02:40	LS 1	66.34	13.42	4.45	15.79			
1.163	04.10.15	02:40	LS 2	75.24	5.33	4.01	15.42			
1.168	04.10.15	05:15	LS 1	67.28	11.52	4.44	16.75			
1.169	04.10.15	05:15	LS 2	71.68	6.06	4.69	17.57			
1.172	04.10.15	09:20	LS 2	73.75	5.54	4.26	16.45			
1.175	04.10.15	13:48	LS 2	71.03	4.18	4.69	20.10			
1.176	04.10.15	16:00	LS 1	67.86	14.49	3.53	14.12			
1.177	04.10.15	16:00	LS 2	74.42	4.28	4.18	17.12			
1.198	08.10.15	19:10	LS 2	83.18	0.63	3.14	13.06			
1.199	08.10.15	19:15	LS 1	75.66	7.53	3.60	13.21			
1.202	09.10.15	00:45	LS 1	72.88	7.54	4.18	15.40			
1.203	09.10.15	00:45	LS 2	80.45	0.83	3.83	14.89			
1.204	09.10.15	04:10	LS 1	72.61	7.11	4.26	16.02			
1.205	09.10.15	04:10	LS 2	79.00	0.42	4.09	16.50			
1.212	09.10.15	15:20	LS 2	78.17	1.15	4.34	16.34	0.11	1.33	0.0041
1.213	09.10.15	15:40	LS 1	72.82	6.58	3.92	16.68			
1.225	10.10.15	08:20	LS 2	81.43	3.34	2.81	12.41			
1.226	10.10.15	08:20	LS 1	74.27	11.42	2.49	11.83			
1.228	10.10.15	12:00	LS 2	79.28	6.06	2.61	12.05			
1.229	10.10.15	12:00	LS 1	75.79	10.46	2.22	11.53			
1.230	10.10.15	15:45	LS 2	81.86	3.76	2.53	11.85			
1.232	10.10.15	16:00	LS 1	76.17	8.58	2.46	12.79	1.10	4.27	0.0172
1.254	11.10.15	17:35	HE$_{calc}$	68.42	13.63	6.01	15.86	3.26	5.99	0.0076
1.261	12.10.15	01:00	Filter$_{carb}$	29.71	27.66	3.48	35.02	7.58	17.48	0.0132
1.267	12.10.15	05:30	Filter$_{calc}$	19.67	27.97	5.12	40.53	11.79	13.65	0.0157
1.268	12.10.15	07:30	LS 1	77.03	7.21	9.97	5.78			
1.269	12.10.15	07:30	LS 2	82.69	1.04	9.45	6.82			
1.271	12.10.15	13:34	LS 1	76.58	7.74	9.81	5.86	1.15	1.33	0.0038

Table A.1: Sample analysis for test campaign 1.

	Date	Time	Spot	CaO wt.%	$CaCO_3$ wt.%	$CaSO_4$ wt.%	Ash wt.%	C wt.%	BET m^2/g	PV cm^3/g
1.272	12.10.15	13:34	LS 2	83.01	0.42	10.30	6.28	0.13	1.71	0.0062
1.276	12.10.15	17:00	LS 2	82.84	0.52	10.63	6.01	0.12	1.31	0.0038
1.277	12.10.15	17:00	LS 1	76.29	7.95	10.32	5.43	1.11	1.32	0.0036
1.279	12.10.15	19:50	LS 1	75.46	8.99	10.24	5.31			
1.280	12.10.15	19:50	LS 2	82.45	0.10	11.14	6.31			
1.287	13.10.15	21:00	LS 1	77.27	15.73	1.01	5.99			
1.288	13.10.15	21:00	LS 2	91.78	1.46	1.04	5.72			
1.290	14.10.15	00:30	LS 1	78.54	13.74	1.12	6.60			
1.291	14.10.15	00:30	LS 2	91.75	1.04	1.01	6.20	0.15	2.30	0.0084
1.293	14.10.15	04:15	LS 2					0.20	2.19	0.0082

Table A.2: Sample analysis for test campaign 2.

	Date	Time	Spot	CaO wt.%	$CaCO_3$ wt.%	$CaSO_4$ wt.%	Ash wt.%	C wt.%	BET m^2/g	PV cm^3/g
2.5	23.11.15	17:40	LS 1	86.16	8.18	1.11	4.55			
2.6	23.11.15	17:40	LS 2	93.66	0.09	1.20	5.05			
2.17	24.11.15	03:37	LS 2	94.14	0.08	0.95	4.83			
2.18	24.11.15	04:00	LS 1	86.73	7.74	0.89	4.64			
2.21	24.11.15	08:00	LS 1	86.90	7.73	0.92	4.45			
2.22	24.11.15	08:00	LS 2	94.46	0.43	0.82	4.29			
2.28	24.11.15	15:20	LS 1	87.17	8.05	0.63	4.16			
2.29	24.11.15	15:17	LS 2	94.02	0.38	0.86	4.74			
2.32	24.11.15	20:10	LS 1	85.84	9.01	0.57	4.58			
2.33	24.11.15	20:10	LS 2	94.38	0.18	0.62	4.82			
2.45	25.11.15	08:45	LS 2	94.41	0.14	0.61	4.84			
2.48	25.11.15	09:05	LS 1	87.16	7.91	0.61	4.32			
2.51	25.11.15	12:50	LS 2	94.84	0.09	0.59	4.48			
2.52	25.11.15	13:15	LS 1	87.43	7.90	0.51	4.16			
2.54	25.11.15	16:50	LS 2	94.57	0.07	0.55	4.80			
2.55	25.11.15	16:50	LS 1	86.79	8.29	0.51	4.41			
2.58	25.11.15	21:06	LS 1	87.33	8.05	0.45	4.18			
2.59	25.11.15	21:06	LS 2	94.23	0.25	0.56	4.97			
2.65	25.11.15	21:06	LS 1	85.59	7.95	0.47	6.00	1.09		

Table A.2: Sample analysis for test campaign 2.

	Date	Time	Spot	CaO wt.%	CaCO$_3$ wt.%	CaSO$_4$ wt.%	Ash wt.%	C wt.%	BET m^2/g	PV cm^3/g
2.78	27.11.15	12:45	LS 2	93.33	1.18	0.75	4.73			
2.79	27.11.15	12:45	LS 1	87.05	5.81	0.84	6.29			
2.86	27.11.15	17:05	HE$_{carb}$	83.75	8.11	1.42	8.14	1.10	1.13	0.0022
2.87	27.11.15	17:15	HE$_{calc}$	72.36	15.43	4.70	12.21	1.94	0.96	0.0023
2.88	27.11.15	20:00	LS 1	86.91	6.59	1.21	5.29	0.95	0.64	0.0015
2.89	27.11.15	20:00	LS 2	92.07	0.42	1.15	6.35			
2.90	27.11.15	20:00	Filter$_{calc}$	65.23	15.40	7.34	19.37	2.15	1.62	0.0031
2.91	27.11.15	20:00	Filter$_{carb}$	79.92	11.18	1.81	8.90	1.52	1.31	0.0028
2.100	28.11.15	07:45	LS 1	86.25	6.66	1.32	5.78	0.95	0.43	0.0010
2.101	28.11.15	07:45	LS 2	92.11	0.35	1.43	6.12	0.08	0.47	0.0011
2.114	28.11.15	20:00	LS 1	85.70	6.27	1.32	6.70			
2.115	28.11.15	20:00	LS 2	91.51	0.95	1.28	6.26			
2.118	29.11.15	00:20	LS 1	86.47	6.17	1.33	6.03	0.88	0.45	0.0010
2.119	29.11.15	00:20	LS 2	91.91	0.35	1.39	6.35	0.09	0.47	0.0012
2.122	29.11.15	04:05	LS 1	86.07	6.29	1.43	6.22			
2.123	29.11.15	04:05	LS 2	90.74	1.54	1.41	6.30			
2.126	29.11.15	08:00	LS 1	85.34	6.16	1.58	6.92			
2.127	29.11.15	08:00	LS 2	91.41	0.94	1.49	6.17	0.13	0.55	0.0014
2.134	29.11.15	08:45	LS 1	89.42	0.00	1.62	8.96	0.95		
2.143	30.11.15	00:15	LS 2	89.42	2.72	1.22	6.64			
2.144	30.11.15	00:15	LS 1	84.84	8.32	1.14	5.70			
2.151	30.11.15	07:30	LS 1	84.38	8.39	1.04	6.19			
2.152	30.11.15	07:30	LS 2	91.29	1.13	1.15	6.42			
2.161	30.11.15	06:30	LS 2	91.58	0.75	0.98	6.68			
2.162	30.11.15	16:35	LS 1	84.07	9.47	0.86	5.60			
2.163	30.11.15	20:00	LS 1	83.70	9.35	0.90	6.05			
2.164	30.11.15	20:00	LS 2	91.13	2.02	0.99	5.86			
2.167	01.12.15	00:11	LS 1	84.86	9.26	0.75	5.13			
2.168	01.12.15	00:11	LS 2	92.06	0.41	0.89	6.65			
2.172	01.12.15	04:00	LS 1	85.47	9.22	0.68	4.63			
2.173	01.12.15	04:00	LS 2	92.65	1.66	0.89	4.80	0.22	0.73	0.0017
2.200	02.12.15	19:50	LS 1	86.22	9.62	0.31	3.85			
2.201	02.12.15	19:50	LS 2	94.48	0.30	0.40	4.82			
2.205	03.12.15	01:20	LS 2	94.10	0.98	0.44	4.49			

Table A.2: Sample analysis for test campaign 2.

	Date	Time	Spot	CaO wt.%	CaCO$_3$ wt.%	CaSO$_4$ wt.%	Ash wt.%	C wt.%	BET m^2/g	PV cm^3/g
2.206	03.12.15	01:20	LS 1	85.66	10.07	0.25	4.02			
2.207	03.12.15	04:30	LS 1	86.25	9.66	0.24	3.85			
2.208	03.12.15	04:30	LS 2	95.15	0.62	0.25	3.98	0.06	0.79	0.0019

Table A.3: Sample analysis for test campaign 3.

	Date	Time	Spot	CaO wt.%	CaCO$_3$ wt.%	CaSO$_4$ wt.%	Ash wt.%	C wt.%	BET m^2/g	PV cm^3/g
3.7	21.02.16	04:45	LS 1	82.22	8.70	3.63	5.45			
3.8	21.02.16	04:45	LS 2	70.29	23.15	2.31	4.25			
3.14	21.02.16	20:15	LS 2	88.86	2.13	2.19	6.83			
3.15	21.02.16	20:15	LS 1	86.53	4.78	2.05	6.64			
3.19	22.02.16	05:00	LS 1	82.71	5.73	4.16	7.40			
3.20	22.02.16	05:00	LS 2	86.73	1.15	4.47	7.65			
3.24	22.02.16	16:00	LS 1	81.96	5.24	5.33	7.47			
3.25	22.02.16	16:00	LS 2	80.55	5.75	5.43	8.28			
3.28	22.02.16	19:45	LS 1	81.00	5.06	5.38	8.57			
3.29	22.02.16	19:45	LS 2	84.44	1.44	5.50	8.62			
3.31	23.02.16	01:35	LS 1	80.41	5.18	5.09	9.32			
3.32	23.02.16	01:35	LS 2	84.00	1.82	5.54	8.64			
3.34	23.02.16	04:49	LS 1	82.56	4.74	3.57	9.13			
3.35	23.02.16	04:49	LS 2	82.00	5.91	3.73	8.36			
3.37	23.02.16	09:00	LS 1	83.22	5.31	2.31	9.16			
3.38	23.02.16	09:00	LS 2	85.75	2.13	3.07	9.05			
3.40	23.02.16	12:50	LS 1	83.87	5.16	2.13	8.84			
3.41	23.02.16	12:55	LS 2	86.65	2.00	2.32	9.03	0.20	0.80	0.0019
3.42	23.02.16	17:16	LS 2	82.87	5.37	2.14	9.62			
3.43	23.02.16	17:30	LS 1	83.84	4.91	1.53	9.72			
3.46	23.02.16	17:30	HE$_{carb}$					3.16		
3.47	23.02.16	17:30	HE$_{calc}$	20.75	27.52	14.78	51.74	5.87		
3.48	25.02.16	04:04	LS 1	85.46	3.47	2.38	8.69			
3.49	25.02.16	04:04	LS 2	84.10	4.97	2.66	8.27			
3.52	25.02.16	07:52	LS 1	84.02	4.89	1.63	9.46			
3.53	25.02.16	07:52	LS 2	86.09	3.22	2.32	8.38			

Table A.3: Sample analysis for test campaign 3.

	Date	Time	Spot	CaO wt.%	CaCO$_3$ wt.%	CaSO$_4$ wt.%	Ash wt.%	C wt.%	BET m^2/g	PV cm^3/g
3.58	25.02.16	17:00	LS 1	84.56	5.14	1.06	9.24			
3.59	25.02.16	17:00	LS 2	86.02	0.79	1.62	11.56			
3.65	26.02.16	00:01	LS 1	83.84	4.91	1.89	9.35			
3.66	26.02.16	00:01	LS 2	88.10	0.56	1.64	9.70			
3.71	26.02.16	08:00	LS 1	83.86	5.60	1.46	9.08			
3.72	26.02.16	08:00	LS 2	85.89	0.67	1.75	11.69			
3.74	26.02.16	12:00	LS 1	84.18	5.25	1.40	9.18			
3.75	26.02.16	12:00	LS 2	87.50	0.90	1.65	9.96			
3.81	26.02.16	20:40	LS 1	84.11	4.39	2.25	9.26			
3.82	26.02.16	20:40	LS 2	89.54	0.15	1.76	8.56			
3.83	27.02.16	00:01	LS 1	85.10	4.34	2.34	8.22			
3.84	27.02.16	00:01	LS 2	88.83	0.02	2.27	8.88			
3.85	27.02.16	00:00	Bett$_{carb}$	81.09	7.69	2.11	11.22	1.10		
3.86	29.02.16	00:04	Filter$_{carb}$					1.58		
3.87	27.02.16	00:00	Filter$_{carb}$					18.71		
3.88	27.02.16	05:00	LS 1	82.95	5.41	1.42	10.22			
3.89	27.02.16	05:00	LS 2	87.75	0.50	1.98	9.77			
3.93	27.02.16	07:10	LS 1	81.86	5.02	2.00	11.12			
3.94	27.02.16	07:10	LS 2	86.01	0.56	2.04	11.38			
3.96	27.02.16	09:50	LS 1	80.55	5.20	2.28	11.96			
3.97	27.02.16	09:50	LS 2	85.54	1.27	2.15	11.04			
3.99	27.02.16	16:30	LS 1	81.07	7.82	1.91	9.20			
3.100	27.02.16	16:30	LS 2	83.82	0.54	2.50	13.14			
3.104	28.02.16	00:01	LS 1	79.16	8.77	2.08	9.99			
3.105	28.02.16	00:01	LS 2	86.20	2.36	2.23	9.20			
3.112	28.02.16	09:00	LS 1	81.81	9.02	1.28	7.89			
3.113	28.02.16	09:00	LS 2	83.84	2.92	1.90	11.34			
3.119	28.02.16	16:50	LS 1	80.78	8.68	0.99	9.55			
3.120	28.02.16	16:50	LS 2	89.32	0.83	1.24	8.60			
3.121	28.02.16	20:25	LS 1	78.95	9.61	0.92	10.52			
3.122	28.02.16	20:25	LS 2	87.28	2.36	1.17	9.19			
3.124	29.02.16	00:01	LS 1	80.90	9.52	0.90	8.68			
3.125	29.02.16	00:01	LS 2	86.85	1.52	1.07	10.56			
3.127	29.02.16	02:15	LS 1	80.68	9.10	0.83	9.39			

Table A.3: Sample analysis for test campaign 3.

	Date	Time	Spot	CaO wt.%	CaCO$_3$ wt.%	CaSO$_4$ wt.%	Ash wt.%	C wt.%	BET m^2/g	PV cm^3/g
3.128	29.02.16	02:15	LS 2	89.40	1.36	0.96	8.29			
3.130	29.02.16	04:00	LS 1	80.73	9.19	1.10	8.98			
3.131	29.02.16	04:00	LS 2	88.75	1.06	1.20	8.99			
3.132	29.02.16	00:03	Filter$_{calc}$	45.07	0.00	14.52	54.93	25.72		
3.133	29.02.16	00:04	Filter$_{carb}$	37.33	29.06	2.65	33.61	11.45		
3.134	29.02.16	00:04	Bett$_{carb}$	78.34	12.61	0.86	9.05	1.73		
3.135	29.02.16	08:00	LS 1	81.21	8.60	2.44	7.75			
3.136	29.02.16	08:00	LS 2	85.31	3.28	2.35	9.06			

Table A.4: Sample analysis for test campaign 4.

	Date	Time	Spot	CaO wt.%	CaCO$_3$ wt.%	CaSO$_4$ wt.%	Ash wt.%	C wt.%	BET m^2/g	PV cm^3/g
4.2	02.04.16	18:50	LS 1	87.62	8.62	0.71	3.05			
4.3	02.04.16	18:50	LS 2	90.15	6.02	0.69	3.14			
4.4	03.04.16	01:00	LS 1	88.37	7.11	0.75	3.77			
4.5	03.04.16	01:00	LS 2	93.40	2.84	0.72	3.05			
4.11	03.04.16	12:00	LS 1	87.79	7.80	0.49	3.91			
4.12	03.04.16	12:00	LS 2	94.19	0.33	0.54	4.93			
4.14	03.04.16	16:50	LS 1	88.32	6.75	0.51	4.41			
4.15	03.04.16	16:50	LS 2	94.57	0.35	0.53	4.54			
4.16	03.04.16	20:00	LS 1	87.58	6.88	0.53	5.01			
4.17	03.04.16	20:00	LS 2	93.94	0.90	0.52	4.64			
4.23	08.04.16	04:00	LS 1	87.40	9.17	0.64	2.80			
4.24	08.04.16	04:00	LS 2	94.65	1.40	0.72	3.22			
4.30	08.04.16	16:00	LS 1	87.69	7.51	0.74	4.06			
4.31	08.04.16	16:00	LS 2	94.51	0.27	0.80	4.42			
4.39	08.04.16	20:30	LS 1	87.60	6.77	1.11	4.51			
4.40	08.04.16	20:30	LS 2	90.27	4.89	0.84	4.00			
4.47	10.04.16	08:00	LS 1	88.34	6.25	0.77	4.65			
4.48	10.04.16	08:00	LS 2	93.96	0.42	0.81	4.81			
4.56	10.04.16	19:55	LS 1	87.79	6.96	0.73	4.52			
4.57	10.04.16	19:55	LS 2	93.39	0.21	0.77	5.64			
4.59	10.04.16	21:20	HE$_{calc}$	32.13	34.00	19.08	14.78	5.14		

Table A.4: Sample analysis for test campaign 4.

	Date	Time	Spot	CaO wt.%	CaCO$_3$ wt.%	CaSO$_4$ wt.%	Ash wt.%	C wt.%	BET m^2/g	PV cm^3/g
4.66	11.04.16	04:00	LS 1	88.05	7.19	0.82	3.94			
4.67	11.04.16	04:00	LS 2	93.54	0.17	0.91	5.39			
4.70	11.04.16	04:00	Filter$_{calc}$	42.59	13.23	22.34	21.85	2.38		
4.71	11.04.16	07:33	Filter$_{calc}$	50.20	14.38	17.14	18.28	2.49		
4.72	11.04.16	08:30	LS 1	87.10	7.42	0.79	4.69			
4.73	11.04.16	08:30	LS 2	93.74	0.31	0.86	5.08			
4.78	11.04.16	16:00	LS 1	87.42	6.77	0.78	5.03			
4.79	11.04.16	16:00	LS 2	93.86	0.35	0.71	5.08			
4.81	11.04.16	20:00	LS 1	87.34	7.55	0.59	4.52			
4.82	11.04.16	20:00	LS 2	93.04	0.52	0.69	5.74			
4.89	12.04.16	08:00	LS 1	86.65	7.36	0.65	5.34			
4.90	12.04.16	08:00	LS 2	93.95	0.58	0.62	4.84			
4.95	12.04.16	16:00	LS 1	87.38	7.55	0.64	4.43			
4.96	12.04.16	16:00	LS 2	93.96	0.25	0.64	5.15			
4.97	12.04.16	16:00	Bett$_{carb}$	84.60	10.43	0.54	4.43	1.36		
4.98	12.04.16	18:00	LS 1	86.95	8.20	0.57	4.28			
4.99	12.04.16	18:00	LS 2	93.77	0.27	0.65	5.30			
4.105	23.04.16	04:00	LS 1	87.82	7.82	0.64	3.71			
4.106	23.04.16	04:00	LS 2	93.26	1.63	0.69	4.43			
4.114	23.04.16	20:00	LS 1	87.10	8.18	0.43	4.29			
4.115	23.04.16	20:00	LS 2	93.79	0.46	0.55	5.21			
4.123	24.04.16	15:00	Filter$_{carb}$	77.45	14.53	0.95	7.07	1.95		
4.124	24.04.16	08:00	LS 1	86.92	8.26	0.55	4.27			
4.125	24.04.16	08:00	LS 2	94.45	0.27	0.57	4.71			
4.127	24.04.16	12:05	LS 1	86.93	8.56	0.49	4.02			
4.130	24.04.16	14:00	LS 1	86.83	8.85	0.51	3.81			
4.131	24.04.16	14:00	LS 2	94.24	0.33	0.57	4.85			
4.134	24.04.16	16:00	LS 2	94.41	0.63	0.63	4.33			
4.135	24.04.16	16:00	Bett$_{carb}$	84.48	10.95	0.75	3.82	1.50		
4.136	24.04.16	20:00	LS 1	85.77	9.48	0.97	3.78			
4.137	24.04.16	20:00	LS 2	91.99	2.90	0.85	4.26			
4.139	24.04.16	23:40	LS 1	84.19	10.91	1.26	3.64			
4.140	24.04.16	23:40	LS 2	89.53	5.16	1.27	4.04			
4.141	24.04.16	18:30	Filter$_{carb}$	77.73	14.17	1.35	6.74	1.91		

Table A.4: Sample analysis for test campaign 4.

	Date	Time	Spot	CaO wt.%	$CaCO_3$ wt.%	$CaSO_4$ wt.%	Ash wt.%	C wt.%	BET m^2/g	PV cm^3/g
4.142	24.04.16	23:35	Filter$_{calc}$	35.88	19.56	23.06	21.49	3.30		
4.144	25.04.16	04:00	LS 1	84.49	9.84	1.65	4.02			
4.145	25.04.16	04:00	LS 2	89.52	4.89	1.39	4.19			
4.146	25.04.16	04:00	Bett$_{carb}$	80.06	15.71	1.15	3.08	2.14		

Table A.5: Sample analysis for test campaign 5.

	Date	Time	Spot	CaO wt.%	$CaCO_3$ wt.%	$CaSO_4$ wt.%	Ash wt.%	C wt.%	BET m^2/g	PV cm^3/g
5.05	31.10.17	06:30	LS 1	86.37	1.07	3.67	7.32	0.76		
5.12	31.10.17	20:00	LS 4.4	88.50	0.50	4.13	6.22	0.00		
5.25	01.11.17	15:00	LS 1	81.33	7.23	4.62	6.52	0.11		
5.26	01.11.17	15:00	LS 4.4	87.23	0.66	4.59	6.47	0.06		
5.27	01.11.17	16:00	LS 1	81.57	7.00	4.59	6.75	0.05		
5.28	01.11.17	16:00	LS 4.4	87.45	0.64	4.69	6.30	0.00	1.168	
5.29	01.11.17	16:00	Filter$_{calc}$	31.85	23.07	7.91	25.54	9.99		
5.31	01.11.17	16:00	Bett$_{carb}$	79.70	9.77	4.32	5.24	0.22		
5.32	01.11.17	16:00	HE$_{calc}$	64.07	14.05	5.81	13.30	1.64		
5.33	01.11.17	16:00	HE$_{carb}$	69.32	9.73	4.01	14.05	1.27		
5.42	02.11.17	08:00	LS 1	77.91	8.11	7.00	6.66	0.06		
5.43	02.11.17	08:00	LS 4.4	84.48	0.82	7.07	6.34	0.00		
5.51	02.11.17	16:30	LS 1	78.30	8.55	6.78	6.21	0.11		
5.52	02.11.17	16:30	LS 4.4	83.87	1.75	7.04	6.14	0.31		
5.56	03.11.17	00:00	LS 1	77.90	8.82	6.17	6.27	0.03		
5.57	03.11.17	00:00	LS 4.4	85.44	1.07	6.38	6.60	0.00		
5.60	03.11.17	08:30	LS 1	76.89	10.61	5.51	5.76	0.29		
5.61	03.11.17	08:30	LS 4.4	84.62	2.45	5.73	6.73	0.42		
5.67	03.11.17	09:30	LS 1	77.10	10.18	5.71	5.65	0.32		
5.68	03.11.17	09:30	LS 4.4	85.66	1.23	5.95	6.06	0.13	1.252	
5.90	04.11.17	15:30	LS 1	79.67	9.20	3.66	6.44	0.10		
5.91	04.11.17	15:30	LS 4.4	88.50	0.77	3.76	6.78	0.25	1.154	
5.92	04.11.17	15:30	Filter$_{calc}$	55.55	12.09	5.88	16.73	7.76		
5.103	05.11.17	07:00	LS 1	76.02	8.52	8.04	6.60	0.00	1.22	
5.104	05.11.17	07:00	LS 4.4	83.03	0.36	8.40	7.24	0.06	1.33	

Table A.6: Sample analysis for test campaign 6.

	Date	Time	Spot	CaO wt.%	$CaCO_3$ wt.%	$CaSO_4$ wt.%	Ash wt.%	C wt.%	BET m^2/g	PV cm^3/g
6.3	22.11.17	12:00	LS 1	82.14	11.61	0.43	5.12	0.01		
6.20	23.11.17	08:00	LS 1	83.62	10.73	0.29	4.56	0.02		
6.21	23.11.17	08:00	LS 4.4	94.22	0.50	0.32	4.83	0.02		
6.26	23.11.17	14:20	HE_{carb}	82.11	11.80	0.75	4.13	0.21		
6.27	23.11.17	15:45	HE_{calc}	58.58	27.39	6.29	6.03	0.00		
6.30	23.11.17	18:00	LS 1	83.50	10.52	0.27	4.92	0.00		
6.31	23.11.17	18:00	LS 4.4	94.37	0.41	0.29	4.69	0.04	0.86	
6.32	23.11.17	18:00	$Filter_{calc}$	56.72	21.70	10.35	8.23	0.00		
6.33	23.11.17	18:00	$Filter_{carb}$	73.87	19.23	0.87	4.95	0.27		
6.36	24.11.17	00:00	LS 1	82.61	10.41	0.39	5.52	0.07		
6.38	24.11.17	04:00	LS 1	84.16	9.93	0.39	5.02	0.10		
6.49	24.11.17	20:00	LS 1	82.88	11.18	0.29	4.61	0.09		
6.50	24.11.17	20:00	LS 4.4	94.53	0.32	0.34	4.82	0.06		
6.59	25.11.17	12:00	LS 1	84.43	9.91	0.31	4.69	0.00	0.58	
6.60	25.11.17	12:00	LS 4.4	93.58	0.64	0.32	4.72	0.00	0.59	
6.70	01.12.17	21:00	LS 1	86.53	7.07	0.32	5.02	0.00		
6.71	01.12.17	21:00	LS 4.4	94.22	0.45	0.36	4.99	0.06		
6.73	02.12.17	00:00	LS 1	83.78	10.09	0.27	5.09	0.06		
6.74	02.12.17	00:00	LS 4.4	94.21	0.52	0.29	4.87	0.12		
6.93	04.12.17	08:10	LS 4.4	92.99	0.39	0.29	5.08	0.09		
6.94	04.12.17	08:10	LS 1	84.94	8.84	0.26	5.09	0.00		
6.114	05.12.17	15:50	LS 1	83.69	10.20	0.22	4.95	0.00	0.55	
6.115	05.12.17	15:50	LS 4.4	94.12	0.52	0.24	5.13	0.00		
6.118	10.12.17	19:15	LS 1	83.95	8.05	0.26	6.76	0.06		
6.127	11.12.17	12:00	LS 1	84.01	7.84	0.49	7.06	0.05		
6.129	11.12.17	15:36	$Filter_{calc}$	60.41	14.57	11.05	9.46	0.07		
6.131	11.12.17	16:00	LS 1	86.11	8.09	0.17	5.76	0.00		
6.132	11.12.17	16:00	LS 4.4	91.92	0.41	0.20	6.22	0.13		
6.133	11.12.17	16:00	$Filter_{carb}$	75.13	18.48	0.71	5.08	0.14		
6.134	11.12.17	16:00	HE_{carb}	78.31	14.27	0.78	4.92	0.25		
6.135	11.12.17	16:00	HE_{calc}	82.77	5.18	2.84	8.17	0.14		

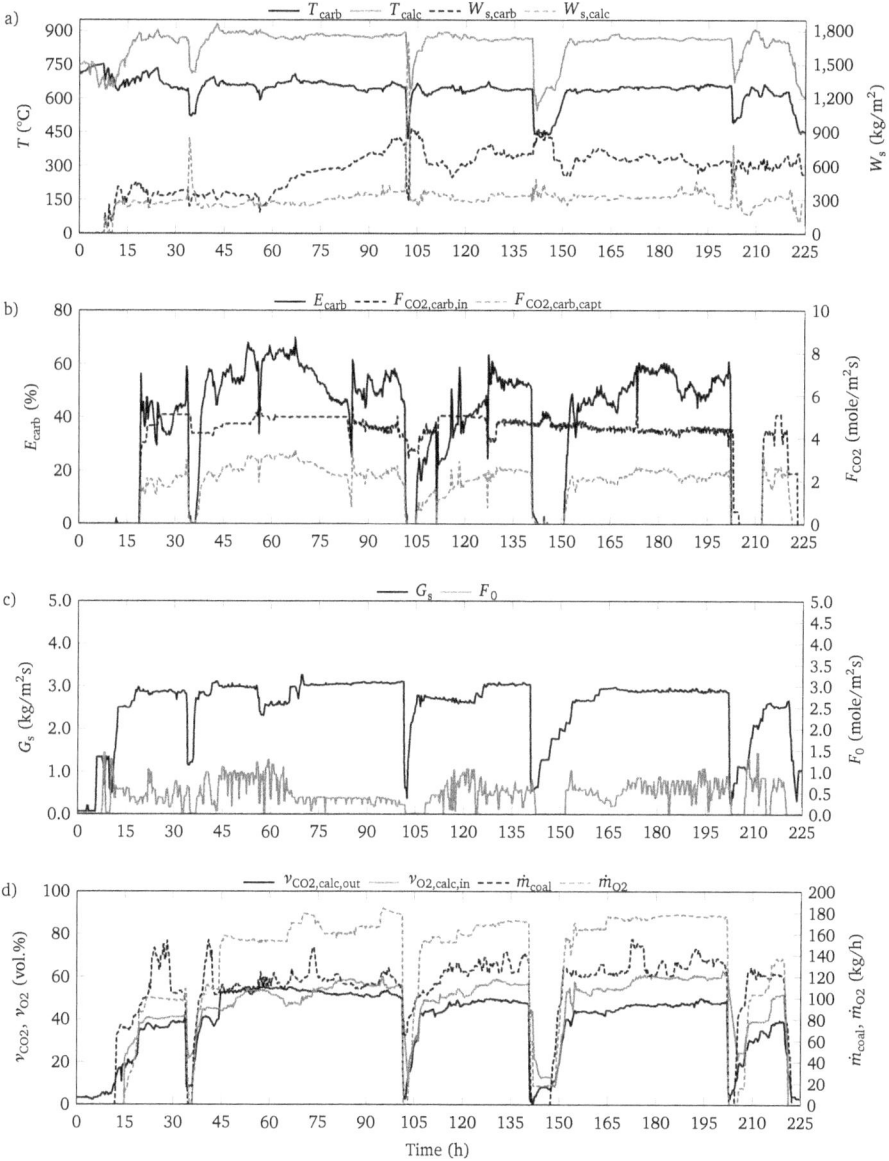

Figure A.1: Test campaign 1 with coarse hard coal - September/October 2015.

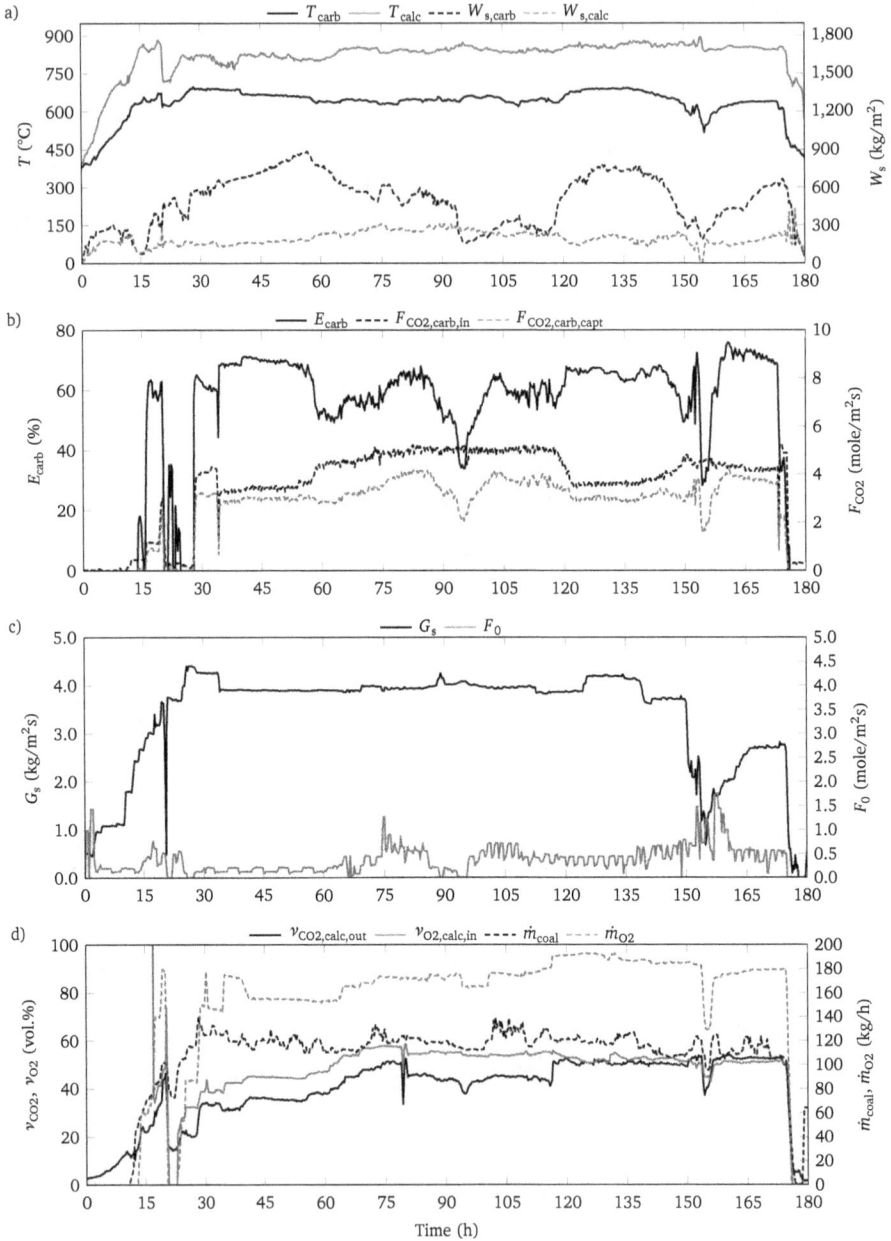

Figure A.2: Test campaign 1 with pulverized hard coal - October 2015.

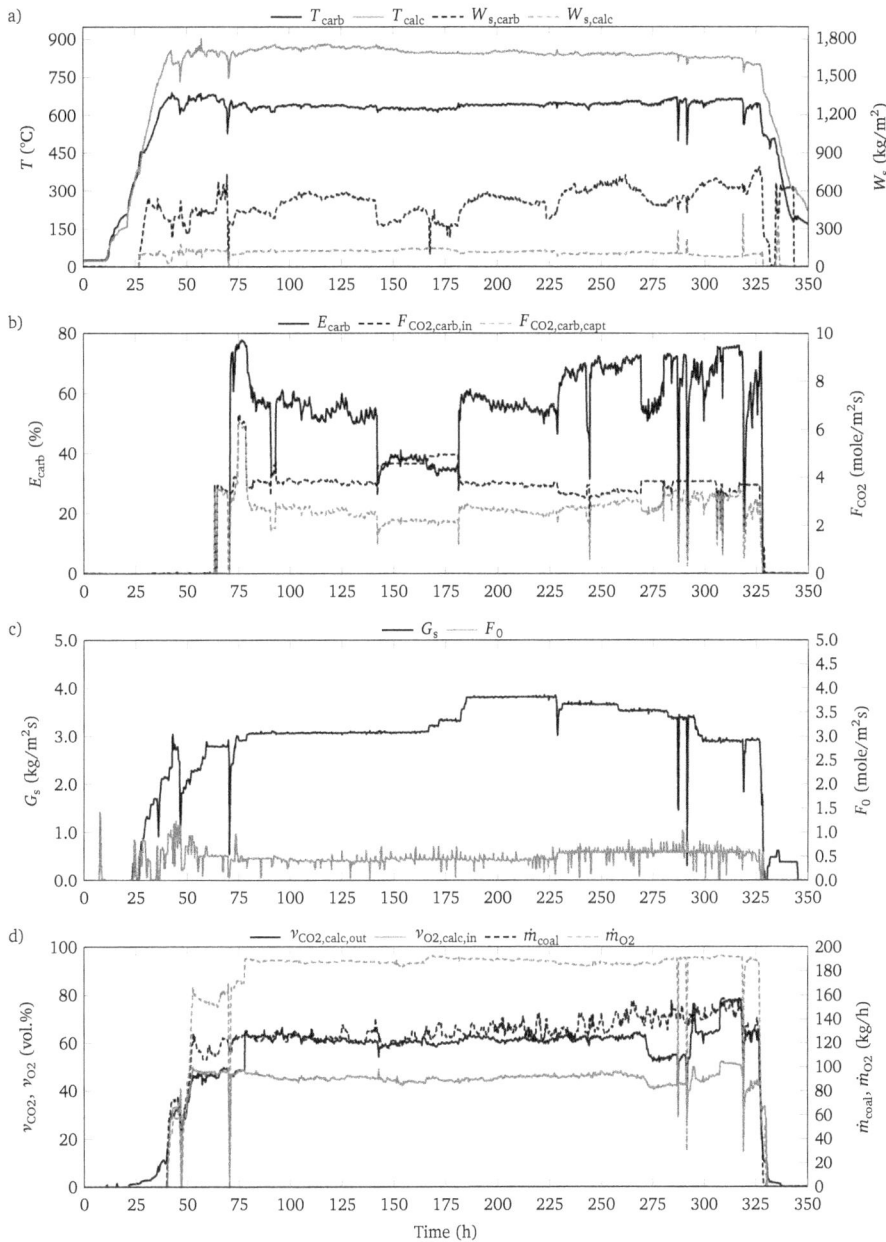

Figure A.3: Test campaign 2 with lignite - November/December 2015.

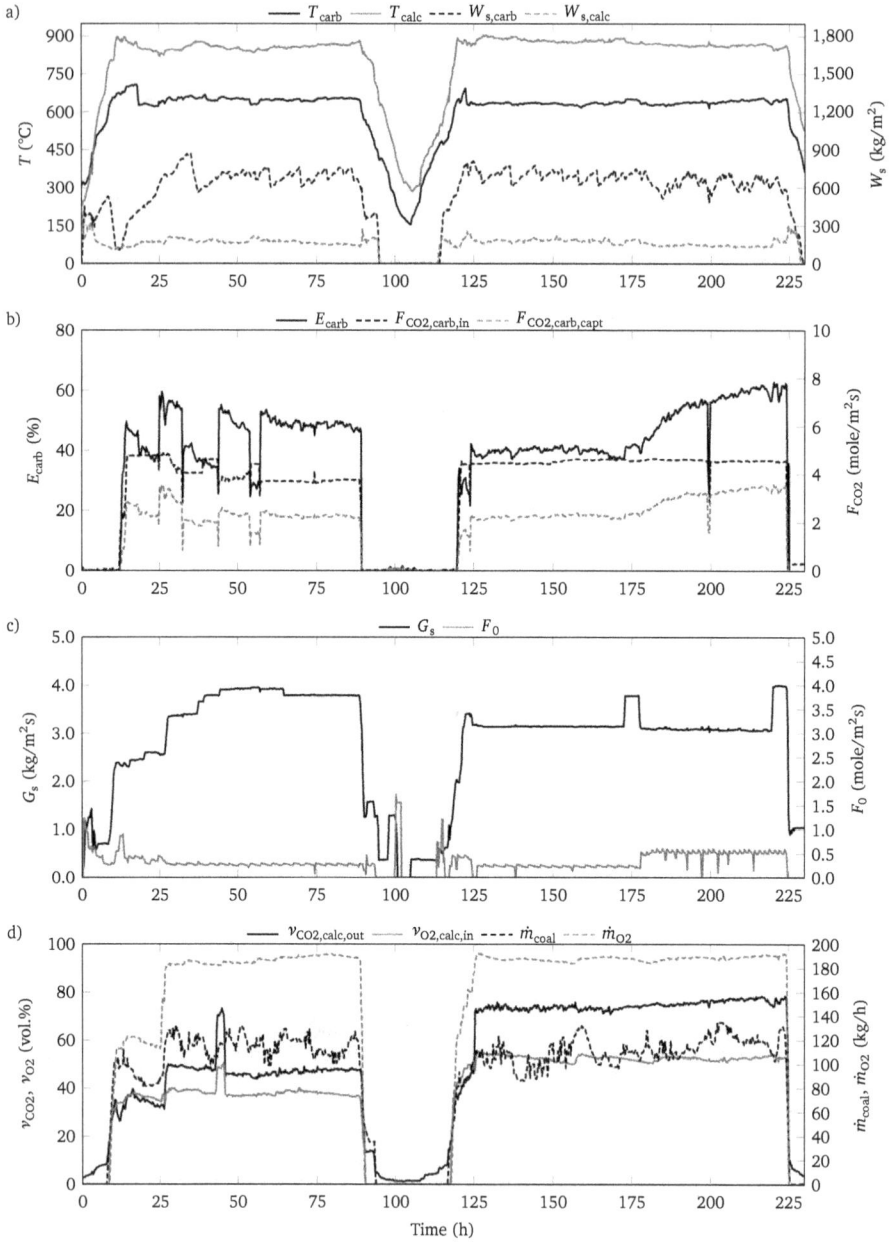

Figure A.4: Test campaign 3 with hard coal - February 2016.

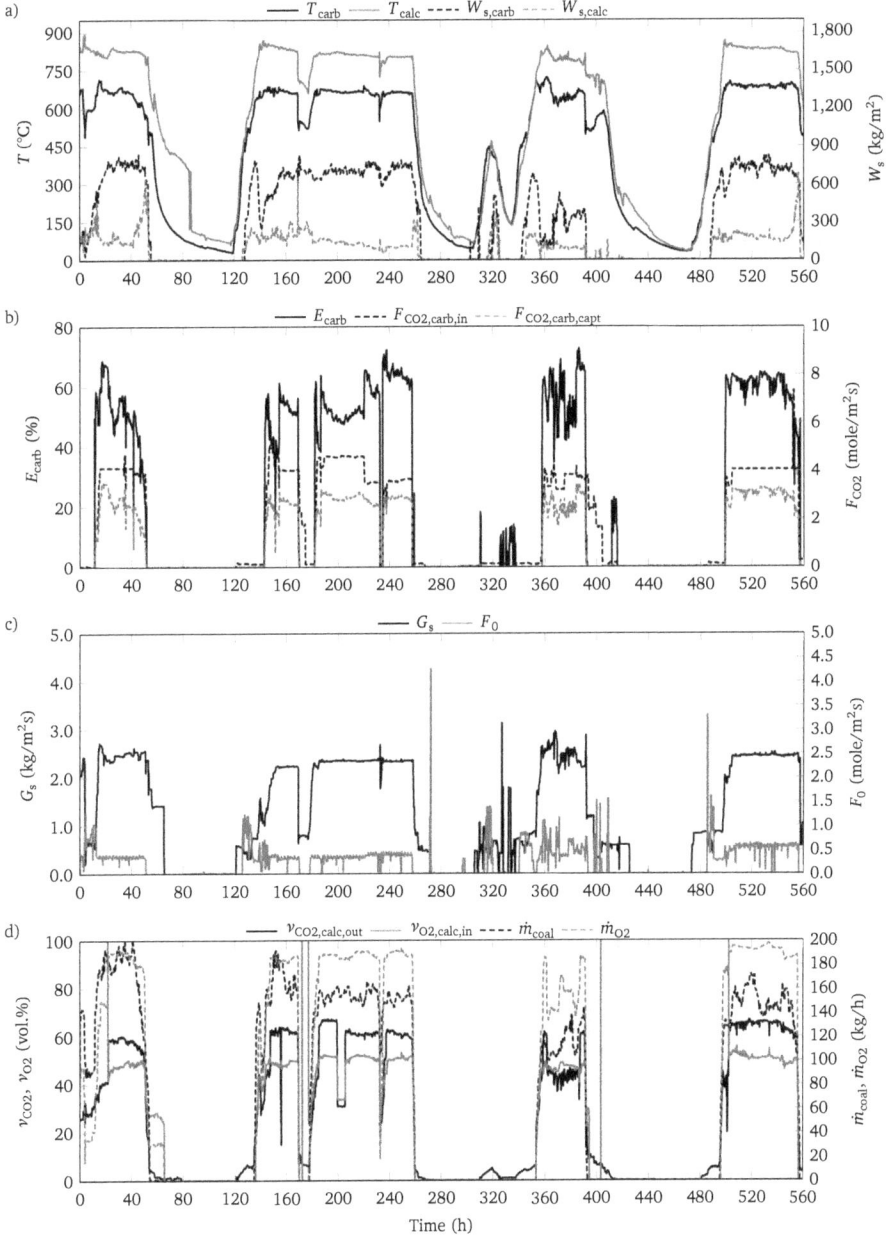

Figure A.5: Test campaign 4 with lignite - April 2016.

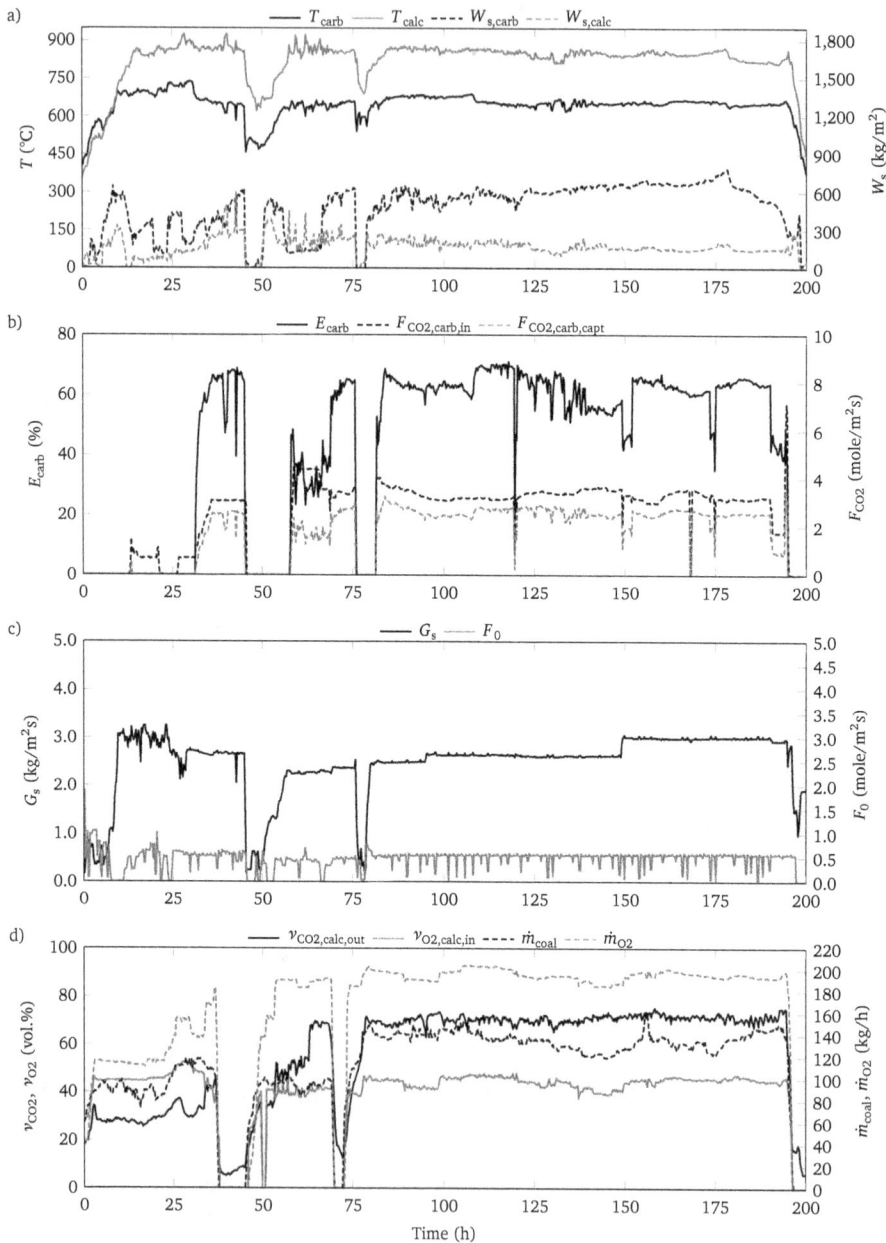

Figure A.6: Test campaign 5 with hard coal - October/November 2017.

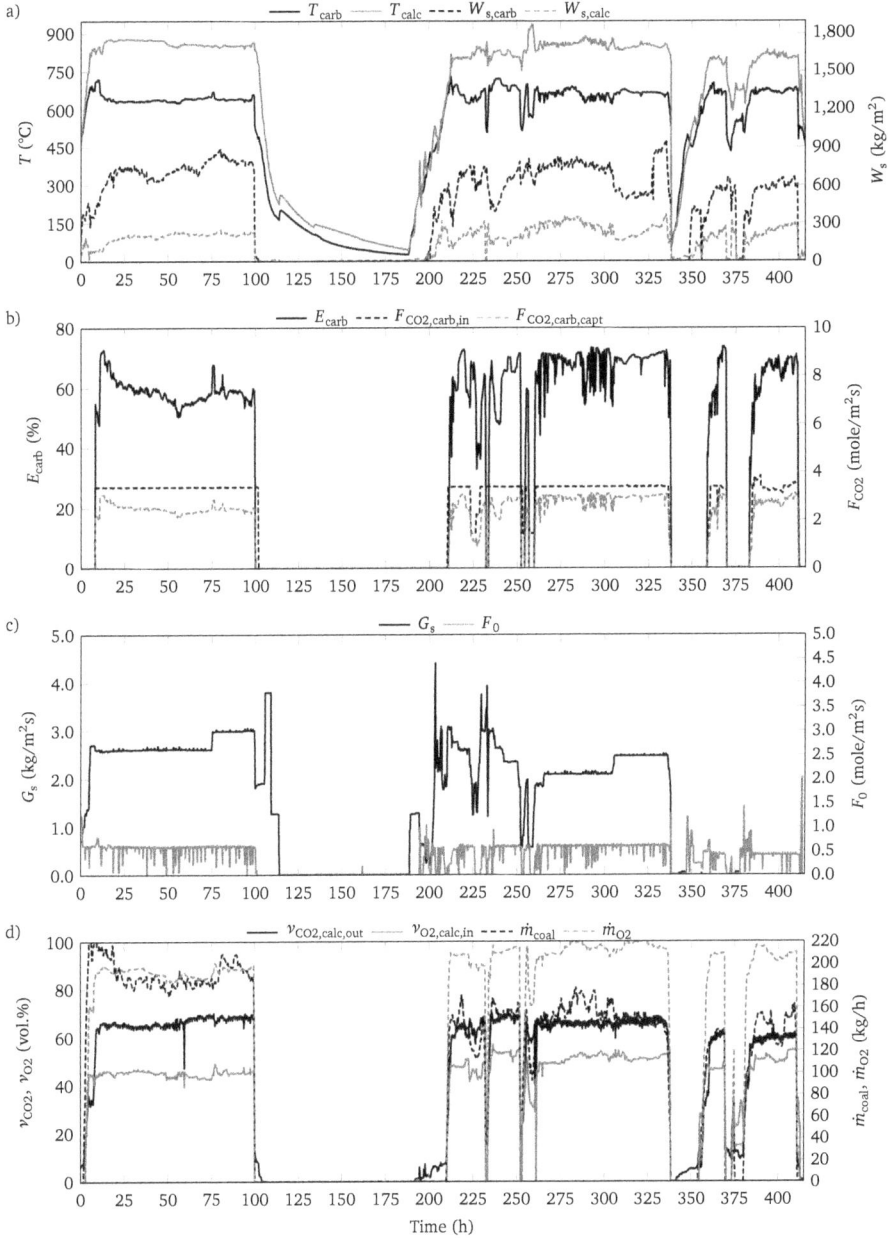

Figure A.7: Test campaign 6 with lignite - November/December 2017.

Journal Publications

2019 **J. Hilz**, M. Haaf, M. Helbig, N. Lindqvist, J. Ströhle, B. Epple, "Scale-up of the carbonate looping process to a 20 MW_{th} pilot plant based on long-term pilot tests," *International Journal of Greenhouse Gas Control*, vol. 88, pp. 332–341, 2019.

2019 M. Haaf, **J. Hilz**, J. Peters, A. Unger, J. Ströhle, B. Epple, "Oxy-Fuel Combustion of Solid Recovered Fuels in the Circulating Fluidized Bed Calciner of a 1 MW_{th} Calcium Looping Unit," *Submitted to Powder Technology*.

2018 M. Haaf, **J. Hilz**, M. Helbig, C. Weingärtner, O. Stallmann, J. Ströhle, B. Epple, "Assessment of the operability of a 20 MW_{th} calcium looping demonstration plant by advanced process modelling," *International Journal of Greenhouse Gas Control*, vol. 75, pp. 224-234, 2018.

2018 **J. Hilz**, M. Helbig, M. Haaf, A. Daikeler, J. Ströhle, and B. Epple, "Investigation of the fuel influence on the carbonate looping process in 1 MW_{th} scale," *Fuel Processing Technology*, vol. 169, pp. 170-177, 2018.

2017 **J. Hilz**, M. Helbig, M. Haaf, A. Daikeler, J. Ströhle, and B. Epple, "Long-term pilot testing of the carbonate looping process in 1 MW_{th} scale," *Fuel*, vol. 210, pp. 892-899, 2017.

2017 M. Helbig, **J. Hilz**, M. Haaf, A. Daikeler, J. Ströhle, and B. Epple, "Long-term Carbonate Looping Testing in a 1 MW_{th} Pilot Plant with Hard Coal and Lignite," *Energy Procedia*, vol. 114, pp. 179-190, 2017.

2017 M. Haaf, A. Stroh, **J. Hilz**, M. Helbig, J. Ströhle, and B. Epple, "Process Modelling of the Calcium Looping Process and Validation Against 1 MW_{th} Pilot Testing," *Energy Procedia*, vol. 114, pp. 167-178, 2017.

2017 **J. Hilz**, M. Haaf, M. Helbig, J. Ströhle, and B. Epple, "Calcium Carbonate Looping: CO_2 capture by using limestone in the cement industry," *Cement International*, vol. 15, pp. 52-63, 2017.

2016 A. Stroh, F. Alobaid, M. T. Hasenzahl, **J. Hilz**, J. Ströhle, and B. Epple, "Comparison of three different CFD methods for dense fluidized beds and validation by a cold flow experiment," *Particuology*, vol. 29, pp. 34-47, 2016.

Scientific contributions to international conferences (presenting author)

2019 *Carbonate Looping for Industrial Application.* DECHEMA-Jahrestreffen der ProcessNet-Fachgruppe Energieverfahrenstechnik und des Arbeitsausschusses Thermische Energiespeicherung, 06.-07.03.2019, Frankfurt am Main, Germany

2018 *Scale-up of the Carbonate Looping Process to a 20 MW_{th} Pilot Plant based on Long-term Pilot Tests.* 14th Conference on Greenhouse Gas Control Technologies (GHGT-14), 22.-26.10.2018, Melbourne, Australia

2018 *Scale-up of Carbonate Looping based on Long-term Pilot Testing in 1 MW_{th} Scale.* DECHEMA-Jahrestreffen der ProcessNet-Fachgruppe Energieverfahrenstechnik, 07.-08.03.2018, Frankfurt am Main, Germany

2017 *1 MW_{th} Long-term Pilot Testing of the Carbonate Looping Process with Hard Coal and Lignite.* 12th International Conference on Fluidized Bed Technology (CFB12), 22.-26.05.2017, Crakow, Poland

2017 *CO_2 Capture with Carbonate Looping for Industrial Application.* 4th International Workshop "CO_2: CCU/CCR and P2X", 15.-16.05.2017, Düsseldorf, Germany

2017 *Long-term Pilot Testing of the Carbonate Looping Process in 1 MW_{th} scale.* IEA Clean Coal Centre's 8th International Conference on Clean Coal Technologies (CCT2017), 08.-12.05.2017, Cagliari, Italy

2017 *Scale-up of Calcium Carbonate Looping Technology for Efficient CO_2 Capture.* CLUSTER Workshop, 27. 04. 2017, Berlin, Germany

2017 *CO_2 capture by means of limestone - Calcium Carbonate Looping.* DECHEMA-Jahrestreffen der ProcessNet-Fachgruppen Abfallbehandlung und Wertstoffrückgewinnung, Energieverfahrenstechnik, Gasreinigung, Hochtemperaturtechnik, Rohstoffe und Kreislaufwirtschaft, 21.-23.03.2017, Frankfurt am Main, Germany

2016 *CO_2 capture using lime.* 3rd Alternative Fuels Symposium, 12.-13.10.2016, Duisburg, Germany

2015 *1 MW_{th} pilot testing and scale-up of the carbonate looping process in the SCARLET project.* 3rd IEAGHG Post Combustion Capture Conference (PCCC3), 08.-11.09.2015, Regina, Canada